Adaptive Hierarchical Isogeometric Finite Element Methods

W0037111

Anh-Vu Vuong

Adaptive Hierarchical Isogeometric Finite Element Methods

Preface by Prof. Dr. Bernd Simeon

 Springer Spektrum   **RESEARCH**

Anh-Vu Vuong
Kaiserslautern, Germany

Vollständiger Abdruck der von der Fakultät für Mathematik der Technischen Universität München zur Erlangung des akademischen Grades eines Doktors der Naturwissenschaften (Dr. rer. nat.) genehmigten Dissertation.

ISBN 978-3-8348-2444-8   ISBN 978-3-8348-2445-5 (eBook)
DOI 10.1007/978-3-8348-2445-5

The Deutsche Nationalbibliothek lists this publication in the Deutsche Nationalbibliografie; detailed bibliographic data are available in the Internet at http://dnb.d-nb.de.

Springer Spektrum
© Vieweg+Teubner Verlag | Springer Fachmedien Wiesbaden 2012

Cover design: KünkelLopka GmbH, Heidelberg

Printed on acid-free paper

Springer Spektrum is a brand of Springer DE. Springer DE is part of Springer Science+Business Media.
www.springer-spektrum.de

Preface

The present monograph is devoted to merge two so far disjoint disciplines, namely geometry and simulation. While geometry and its descendant Computer Aided Geometric Design (CAGD) are concerned with questions of size, shape and the properties of space, numerical simulation deals with discretization methods and their efficient implementation. Intrinsic geometric structures - such as the curved surface of an object or the symmetries and conservation laws of a continuum model - are, in general, not preserved but only approximated in today's simulation software.

If we look closer at the state-of-the-art and select the Finite Element Method (FEM) as well-established technology, we observe that there still exists a great divide between the CAGD approaches for modeling complex geometries and the numerical methods. In particular, the FEM is only able to perform simulations with approximations of free-form shapes. This divide - which can be traced back to the early days of the development in CAGD and for FEM-based numerical simulation - constitutes a severe bottleneck, both in the product development process and in the enhancement of the relevant mathematical technology.

Simulation methods that exactly represent engineering shapes and that vastly simplify the mesh generation by eliminating the necessity to communicate with the CAGD description hold thus great promise. The idea of bridging the gap between the CAGD and the FEM approaches has gained significant momentum with the advent of isogeometric analysis, proposed by T.J.R. Hughes and co-workers in 2005. The present work breaks new ground in the field as it introduces an approach for adaptive refinement based on hierarchical B-splines. In this way, the severe limitations of tensor-product constructions can be overcome, and approximation errors can be controlled. Besides an extensive discussion of the algorithmic aspects, a detailed investigation of the theoretical underpinnings is given, and various numerical test cases demonstrate the potential benefit of an isogeometric simulation.

Anh-Vu Vuong's outstanding dissertation is of far-reaching interest to mathematicians as well as engineers and computer scientists, and I wish him great success with this publication at Springer-Spektrum.

Prof. Dr. Bernd Simeon
Felix-Klein Zentrum für Mathematik
Technische Universität Kaiserslautern

Acknowledgment

Keine Schuld ist dringender,
als die, Dank zu sagen.

(Cicero)

In den vergangenen Jahren haben viele Menschen zum Gelingen dieser Arbeit beigetragen und sollen daher auch nicht unerwähnt bleiben.

Ich danke Herrn Prof. Bernd Simeon für die Betreuung dieser Arbeit und für die Möglichkeit, am Forschungsprojekt EXCITING teilzunehmen. Er hat mich stets in allen meinen Belangen unterstützt und mir dabei auch die nötige Freiheit und Verantwortung überlassen. Ich habe die Freude, auf erfüllte und lehrreiche Doktorandenjahre zurückblicken zu dürfen. Herrn Prof. Bert Jüttler danke ich für die Übernahme des externen Gutachtens und die lebhaften Diskussionen, die stets meinen Horizont um einige geometrische Blickwinkel erweitert haben. Herrn Prof. Peter Rentrop gilt ebenfalls mein Dank für die Übernahme eines Gutachtens und die angenehme Zeit, die ich an seinem Lehrstuhl für Numerische Mathematik verbracht habe.

Meinem Kollegen Christoph Heinrich danke ich für die erfreuliche und produktive Zusammenarbeit und die sehr angenehme Gesellschaft auf Reisen, wo auch immer uns „unser" Projekt hingeführt hatte. Ich danke außerdem meinen anderen Kollegen am Lehrstuhl, insbesondere Martina Pospiech, Florian Augustin, Klaus-Dieter Reinsch, Tobias Weigl und meinem langjährigen Bürogenossen Michael Dörfel. Die kollegiale Atmosphäre und der interne Zusammenhalt haben das universitäre Leben immer wieder bereichert. Über Lehrstuhlgrenzen hinweg gebührt Stefanie Schraufstetter Dank für unterhaltsame und aufschlussreiche Diskussionen bei Keksen und die damit verbundene Ablenkung.

Dank gebührt ebenfalls meinen Studenten, die entweder durch ihre Tätigkeit als studentische Hilfskraft oder in einer von mir betreuten Projekt- oder Abschlussarbeit mit mir in Kontakt standen. Ich hoffe, dass sie in ihrer Zeit bei mir ebensoviel gelernt haben, wie ich von ihnen.

Meinen EXCITING Kollegen danke ich für ihre Gastfreundschaft und den Gedankenaustausch über Länder- und Fachgrenzen hinweg.

Für die sorfältige Durchsicht des Manuskripts bedanke ich mich bei meinem Bruder Anh-Tu Vuong, Tilman Küstner und Julia Niemeyer. Ihr danke ich darüber hinaus für den niemals endenden moralischen Beistand und ihre Hingabe, die mir halfen, auch die Täler des Entstehungsprozesses dieser Arbeit zu durchqueren. Letztendlich bedanke ich mich ganz herzlich bei meiner Familie und ganz besonders bei meinen Eltern für ihre langjährige Unterstützung.

Financial support by the European Union within the 7th Framework Programme, project SCP8-218536 "EXCITING" is gratefully acknowledged.

Garching bei München   Anh-Vu Vuong

Contents

List of Figures

List of Tables

List of Algorithms

Chapter 1
Introduction

Numerical simulation has developed to a key technology in scientific and industrial applications. Still, despite the growing capabilities of modern computers, they are not able to cope with increasing demands of numerical simulation of real-life problems, that raise in the same manner. The limitations are always found in memory or time, when the models get more and more complex and large-scale. It is one of the tasks of numerical analysis to contribute from analytical as well as from algorithmic point of view, that the resources in simulation are used efficiently. One of the most prominent methods for solving partial differential equations is the finite element method (FEM). It is very versatile and can be applied to a wide range of problems on different computational domains. Local refinement and adaptivity for finite elements have been and still are research topics for decades and remarkable progress was made.

In industrial use, however, there is also another bottleneck. The flexibility of finite elements with respect to geometry is based on the use of unstructured meshes consisting of tetrahedral or hexagonal elements. Nevertheless, this comes at the cost of mesh generation, which accounts for a lot of computational and economical resources of the finite element simulation. Especially meshes of good quality in the sense of regularity and with respect to complex geometries are not easily generated without human interaction. Typically these meshes are approximations of designed shape stemming from computer aided geometric design (CAGD), which is an important technology extensively used in many applications, including automotive and aerospace industries. The main task in CAGD is to create, modify and represent any type of shapes. This is done based on the idea to represent a shape with a finite number of points. These are on the one hand used to change the shape and on the other hand used to store the information about the shape. Inspired from aspects like smoothness and interaction the development in this field lead to another description than used for simulation and finite element analysis. Typically the former description is based on splines or piecewise polynomial functions that belong to the most prominent choices in interpolation and approximation.

Both simulation and CAGD share the same environment: they are strongly influenced by the interaction between computers, mathematical research and industrial demands. Just recently a technology called isogeometric analysis has emerged that addresses the costly and involving gap between CAGD and simulation. Therefore it combines concepts from CAGD as well as finite elements by using non-uniform rational B-splines as employed in CAGD for describing the geometry as well as basis functions in the Galerkin projection just like in FEM. This has several effects: on the one hand we can solve the partial differential equation directly on the geometric description and on the other hand obtain different, smooth basis functions. Isogeometric analysis was investigated in numerous articles the potential was shown for a broad scope of application fields.

Still, the problem mentioned in the beginning is not resolved yet. Although isogeometric analysis addresses the interaction with CAGD it is also crucial to enhance adaptivity within the simulation. Precisely, although it is not based on a traditional kind of mesh it is of utmost importance to increase the resolution locally in order to obtain better numerical results. Many efforts were made in finite element research towards this topic and adaptive finite elements are well-established. The approach taken in this thesis is to identify common structures as well as the differences between isogeometric analysis and finite element methods, which helps transferring ideas from one method to the other. The fundamental idea of FEM is the finite element itself, which is defined for its own and then related with other ones to create something bigger than itself. Due to the arbitrary connectivity for the elements unstructured meshes, which are able to represent domains of arbitrary shape, can be used. The basis functions defined locally on each element form a global basis through the subdivision. As it will be discussed in detail isogeometric analysis uses another approach to achieve the same. Here the focus is not on an element, but on the geometry as a whole. B-splines or NURBS that are pushed-forward onto the computational domain are used as basis functions. Although they have a small support they originate from the global geometry mapping. Furthermore, due to the higher smoothness the basis functions are more intervened with each other.

Due to its origin isogeometric analysis is mainly influenced by reusing CAGD algorithms in a FEM inspired way, which has turned out to be very successful, but on the other hand it is also important not to neglect the numerical point of view. Many local refinement approaches for isogeometric analysis suffer under two difficulties. They either rely too much on its CAGD roots and neglect that the requirements for numerical simulation may differ. For example, it is essential to preserve linear independence and that an efficient assembly of the system matrices is possible. Furthermore, some approaches get lost within the opposing properties — the regularity of an isogeometric mesh, which is influences by its tensor-product structure, and the corresponding tensor-product bases and the irregularity of a locally refined mesh.

The hierarchical approach presented here will address both of these problems. Starting with hierarchical B-splines from CAGD, we make some modifications to guarantee necessary properties by constructions and employ a suitable mesh structure that serves well for the numerical algorithms. By using a hierarchy we preserve the structure within each level but also have the possibility for local refinement through the interaction of elements between the levels.

Scope of this Thesis

Isogeometric analysis contains concepts and ideas from two different areas. Every article about isogeometric analysis deals with B-Splines, NURBS or finite elements and this thesis is not an exception of this. Nevertheless, it is our aim not only to review the spline or FEM foundations, but follow these elaborations to the new field of isogeometric analysis, combine two different viewpoints in it and gain more insight about this method. This makes it a cumbersome task to give a unique summarizing exposition without tiring repetitions. Judging the global picture more valuable than a strictly linear composition we will give cross-references to other sections. So, one emphasize of this thesis is the interplay between geometry and numerical analysis or in other words splines and finite elements. Therefore we will discuss isogeometric analysis in a structured manner, identify its components and relate them to their counterparts in their parent field. The theoretical aspects as well as the practical aspects will be considered. We follow the concept of scientific computing, where the interplay between mathematics, computer science and applications in engineering and sciences is emphasized. On the one hand we will not give too many details about the implementation but we will discuss the main ideas that need to be considered for crafting an isogeometric solver.

State of the Art

This thesis has been written during the time period where the method called isogeometric analysis has started to gather attention on a broad scope. A lot of advances were made by all kind of different people with different background in a very short period of time. The work in this field is coined by a high pace of innovations and developments and therefore it is not an easy task to pin down what is the current state of the art.

The starting point for isogeometric analysis is the highly cited publication [58] that introduced isogeometric analysis. Preceding related work includes WEB-Splines described in [55] that uses B-splines for FEM. This was also discussed in [60]. The use of NURBS at the boundary is done in NURBS-enhanced FEM (NEFEM) shown in [86]. Soon the isogeometric analysis began to mature further: The convergence theory was initiated in [18] and is continued in [21]. Isogeometric analysis has been applied to numerous fields of applications. The incomplete enumeration includes structural analysis [36], fluid-structure interaction [20], shape optimization [98, 70], contact mechanics [93], shell analysis [63, 23] and problems in electromagnetism [29]. They were investigated with regard of mixed approaches [29], locking free formulations [43, 8] and mesh distortion [69].

For local refinement much less results have been achieved. T-splines were investigated in [42] and [19]. Their properties were critically investigated in [28] and further adopted to the simulation requirements in [83]. Other concepts were also transfered from CAD to isogeometric analysis like HPT-splines shown in [74]. Nevertheless, there is no dominant refinement methods for isogeometric analysis available. The hierarchical refinement concept that will be presented here is also found in [95].

Structure of the Thesis

This thesis is divided into six chapters. The first two chapters after the introductory one resemble the two major cornerstones within isogeometric analysis. **Chapter 2** starts with basic prerequisites from *applied geometry and computer aided geometric design*. Fundamental geometric objects and spaces are introduced, their properties and their algorithmic treatment are discussed. **Chapter 3** gives a short introduction into *mathematical models and their numerical solution with finite element methods*. We discuss fundamental models used in this thesis and derive their variational formulation. The finite element method is introduced, such that it can be compared to the approach of isogeometric analysis. The main purpose of this chapter is to fix the notation for the problem formulation themselves and review the methods well known already. The main part begins in **Chapter 4** where the topics of the previous chapters are merged and lead into *isogeometric analysis*, a method for solving partial differential equation under consideration and usage of a geometric description. The method is discussed in detail, new concepts and viewpoints are developed and some numerical results are shown. In **Chapter 5** we focus on a very special and utmost challenging topic and deal with *adaptive hierarchical local refinement in isogeometric analysis*. An overview of existing techniques and their main ideas are discussed. Then the theoretical and historical background of hierarchical approaches to splines is given. The major topic of this thesis, hierarchical adaptive refinement for isogeometric analysis is investigated thoroughly based on the foundations in Chap. 4 and numerical examples are given. At the end in **Chapter 6** the achievements are summarized and interpreted. Further information about software as well as the geometric data used for the simulation examples can be found in the appendix.

Notational Conventions

Dealing with mathematical modeling, numerical analysis and computational geometry the notation in this work does not fit into the standard notation in each field due to consistency. Nevertheless, some standard notational conventions are valid throughout all chapters:

Mathematical statements as well as equations are numbered consecutively within each chapter. Dependencies of parameters are sometimes omitted if they are clear from the context. Vectors and matrices are printed boldface. The Kronecker delta is defined as

$$\delta_{ij} = \begin{cases} 1, & \text{if } i = j \\ 0, & \text{if } i \neq j \end{cases} \tag{1.1}$$

A subset Ω is always meant to be a bounded domain, i.e. an open connected set. For a given set $K \subset \mathbb{R}^d$ we denote the closure as $\overline{K}$, the interior as $\text{int} K$, the boundary as ∂K and the diameter of K as $\text{diam}(K)$. The power set of K is denoted by $\mathfrak{P}(K)$. The support of a function is defined as

$$\text{supp}(f) := \overline{\{x \mid f(x) = 0\}} \tag{1.2}$$

The elements of a vector $\boldsymbol{x} \in \mathbb{R}^n$ are denoted by $x_i, i = 1, \ldots, n$ and the ones of a matrix $\boldsymbol{A} \in \mathbb{R}^{m \times n}$ as

$$a_{ij} = (\boldsymbol{A})_{ij}, \ i = 1, \ldots, m, \ j = 1, \ldots, n. \tag{1.3}$$

The trace of a quadratic matrix $\boldsymbol{A} \in \mathbb{R}^{n \times n}$ is denoted by

$$\mathrm{tr}(\boldsymbol{A}) = \sum_{i=1}^{n} a_{ii}. \tag{1.4}$$

Partial derivatives are denoted by

$$\frac{\partial}{\partial x_i} f(\boldsymbol{x}) = \partial_i f(\boldsymbol{x}). \tag{1.5}$$

and analogously for multiple partial derivatives. If appropriate multiindices $\alpha = (\alpha_i, \ldots, \alpha_n)$ may be used. The Jacobian matrix of a vector valued function $\boldsymbol{F}$ is denoted by $D\boldsymbol{F}$ or $\nabla \boldsymbol{F}$. We will use the nabla operator for the divergence of a vector field

$$\nabla \cdot \boldsymbol{u} = \sum_i \partial_i u_i. \tag{1.6}$$

The unit sphere in $\mathbb{R}^3$ is denoted by S^2.

If V is a normed linear space, V' is its dual space. $C^k(\Omega)$ denotes the space of functions with k continuous derivatives over Ω, $C^\infty(\Omega)$ the space of arbitrary differentiable functions over the domain Ω and $C_0^\infty(\Omega)$ the subset of $C^\infty(\Omega)$, whose functions have compact support in Ω.

We call $L^2(\Omega)$ the Lebesgue space of square-integrable functions over Ω. Moreover we denote the Sobolev spaces with

$$H^m(\Omega) := \{u \in L^2(\Omega) | D^\alpha u \in L^2(\Omega) \forall |\alpha| \le m\} \tag{1.7}$$

with the weak derivative $D^\alpha u$. We have the Sobolev norm $||u||_{H^m} := (\sum_{|\alpha| \le m} |D^\alpha u||^2)^{\frac{1}{2}}$ and seminorm $|u|_{H^m} := (\sum_{|\alpha| = m} |D^\alpha u||^2)^{\frac{1}{2}}$. The closure of $C_0^\infty(\Omega)$ in $H^m(\Omega)$ with respect to the Sobolev norm is denoted by $H_0^m(\Omega)$. The euclidean product in $\mathbb{R}^d$ or the inner product in $L^2(\Omega)$ is denoted by $(\cdot, \cdot)$.

Chapter 2

Prerequisites from Applied Geometry and Spline Theory

Geometry is not true, it is advantageous.

(Henry Poincare)

In this chapter we will review some basics of applied geometry that are essential prerequisites for the following chapters. In particular, the concepts from computer aided geometric design (CAGD) are mandatory for isogeometric analysis, but we will also look at other prerequisites that in the same manner influence finite element analysis.

Of course it would be a futile attempt to be nearly complete and present the whole complexity of the field. Therefore we will often restrict ourselves to one viewpoint and especially focus on the aspects necessary for the following chapters.

After having introduced parameterizations of geometric objects, we will look at polynomial spline spaces and a special basis, namely B-Splines. We will summarize their properties, which will turn out to be useful in the following chapters. Then also rational splines and NURBS will be introduced. Starting from the univariate case we will also investigate multivariate splines. In the end some important algorithmic aspects for evaluation and refinement will be discussed. The importance of the concepts for isogeometric analysis introduced here will be only shortly mentioned in the remarks and corresponding references will be given.

General introductions into CAGD can be found in [46, 56] or [1, 2]. More specific references will be given at corresponding sections.

2.1 Parametric Curves, Surfaces and Volumes

One of the main tasks of CAGD is to represent a geometric object and there are several possible ways to achieve this. We will look at the most natural form how to design and represent shapes on a computer and are interested in the smoothness of this representation.

Definition 2.1. *Let $\Omega_0 \subset \mathbb{R}^{do}$ and $\Omega \subset \mathbb{R}^d$. We call a surjective mapping*

$$G : \Omega_0 \to \Omega, \ x = G(u) \tag{2.1}$$

with $G_i(u) \in C^r(\Omega_0)$ for $i = 1 \ldots d$, a C^r parameterization.

The function G will often just be called geometry mapping or parameterization. The arguments u are called parameters and Ω_0 the parametric space. We also use u, v, w to denote the components of u.

The continuity in Def. 2.1 is also called parametric continuity, because it is based on the parameterization instead of the shape it describes. The so-called geometric continuity describes how smooth the shape of a geometry looks like, e.g. if the tangents in a point of a curve are parallel. If the geometric continuity is higher than the parametric continuity for a given parameterization, it is possible to reparameterize the shape to get a parameterization with the desired smoothness (see e.g. [46]). Nevertheless, in the following we will always deal with parametric continuity and the parameterization is taken as given.

Definition 2.1 is not very demanding and it is often necessary to have further properties.

Definition 2.2. *A C^1 parameterization $G : \Omega_0 \to \Omega$, is regular at u if $DG(u)$ has full rank. G is called regular if it is regular in every point of Ω_0.*

Regular parameterizations allow us to transform integrals over them: for a regular C^1 parameterization $s : \mathbb{R}^2 \to \mathbb{R}^3$ the surface integral may be transformed

$$\int_\Omega f(x)dS = \int_{\Omega_0} f(s(u,v)) |\frac{\partial s}{\partial u} \times \frac{\partial s}{\partial v}| \, du, \tag{2.2}$$

and for a regular C^1 parameterization $l : \mathbb{R} \to \mathbb{R}^3$ a curve integral results in

$$\int_\Omega f(x)dx = \int_{\Omega_0} f(l(u)) \cdot |\frac{dl}{du}| \, du. \tag{2.3}$$

In the case of equal dimensions, integrals over Ω can be transformed into integrals over Ω_0 under the assumption that G is a diffeomorphism by means of the well-known integration rule

$$\int_\Omega f(x) \, dx = \int_{\Omega_0} f(G(u)) |\det DG| \, du. \tag{2.4}$$

For the other direction, differentiation, the chain rule applied to a differentiable function $T(x,y) := T(G(u,v))$ yields

$$\nabla_{(x,y)} T(x,y) = DF(u,v)^{-T} \cdot \nabla_{(u,v)} T(u,v). \tag{2.5}$$

This can be stated analogously for higher dimensions.

Remark 2.3. *The integration rule Eq. (2.4) is needed in FEM and isogeometric analysis for the computation of the stiffness matrix. Surface and line integrals like in Eq. (2.2),(2.3) have to be calculated when dealing with boundary conditions.*

2.2 Spline Functions

When working with parameterizations it is necessary to represent them in a basis with suitable properties. For CAGD it is important that this representation allows to easily change and interact with the geometry. In this section we will look at spline functions, which fulfill these requirements.

2.2.1 Polynomial Spaces

In order to introduce splines we first need to introduce polynomials.

Definition 2.4. *The polynomial space Π of degree m over the interval $[a, b] \subset \mathbb{R}$ is defined as*

$$\Pi^m([a, b]) := \{p(x) \in \mathcal{C}^\infty([a, b]) \mid p(x) = \sum_{i=0}^{m} c_i x^i, c_i \in \mathbb{R}\}. \tag{2.6}$$

Obviously the monomials are a basis of the polynomial space, but there is another one with more local supports, the Bernstein basis defined as

$$b_{i,p}(x) = \binom{p}{i} x^i (1 - x)^{p-i}. \tag{2.7}$$

Polynomials are the classical choice for solving interpolation problems: let V_n be a subspace with dimension n of the space V and $L_i \in V', i = 1 \ldots n$, linear continuous functionals. With n given values $f_i \in \mathbb{R}, i = 1 \ldots n$ we seek $u_n \in V_n$ such that

$$L_i u_n = f_i, \quad i = 1, \ldots, n. \tag{2.8}$$

Typically the linear functionals are point evaluations

$$L_i : C^m[a, b] \to \mathbb{R}, \ L_i(f) = \frac{d^{m_i}}{dx^{m_i}} f(x_i). \tag{2.9}$$

In the case $m_i = 0$ we call it Lagrange interpolation and otherwise with additional information about derivatives Hermite interpolation.

Remark 2.5. *Lagrange and Hermite interpolation are directly connected to basis functions on elements in FEM (see Sec. 3.4.1). The Bernstein basis is of importance for the concept of Bézier extraction (see Sec. 4.4.2) and as a predecessor for B-splines.*

2.2.2 Spline Spaces

Although polynomials are easy to evaluate, they show an unsatisfying behavior towards interpolation. Piecewise polynomials called splines are a remedy for this. In textbooks about numerical analysis we usually find the twice continuously differentiable cubic spline, which shows superior interpolatory behavior compared to polynomials. Here we will state a more general definition.

Definition 2.6. *Let $a = x_0 < x_1 < \cdots < x_k = b$ be a partition of the interval $[a, b]$, m a positive integer and $m_1, \cdots, m_{k-1}$ integers with $1 \leq m_i \leq p+1$, $i = 1, \ldots, k-1$. We call a function $s : [a, b] \to \mathbb{R}$ a* polynomial spline *of degree p if*

$$s_i := s|_{[x_i, x_{i+1}]} \in \Pi^p([x_i, x_{i+1}]), \quad i = 0, \ldots, k-1 \tag{2.10a}$$

and

$$\frac{d^j}{dx^j} s_{i-1}(x_i) = \frac{d^j}{dx^j} s_i(x_i) \text{ for } j = 0, \cdots, p - m_i \text{ and } i = 1, \ldots, k-1. \tag{2.10b}$$

If $m_i = p+1$ we understand the second condition to allow a discontinuity between the two polynomial pieces.

The main advantage of these functions are clear: they can be evaluated quickly as they consists of polynomials and they are flexible, because of their piecewise definition. Furthermore, the continuity of the function can be prescribed.

Later on we will make use of an alternative description: based on a partition and the multiplicities m_i we call a sequence $u_0 \leq \cdots \leq u_{2p+M}$ with $M = \sum_{i=1}^{k-1} m_i$ an *extended partition* if

$$u_0 = \cdots = u_p = a \text{ and } u_{p+M+1} = \cdots = u_{2p+M+1} = b \tag{2.11a}$$

and

$$(u_{p+1}, \ldots, u_{p+M}) = (\underbrace{x_1, \ldots, x_1}_{m_1 \text{times}}, \ldots, \underbrace{x_i, \ldots, x_i}_{m_i \text{times}} \ldots, \underbrace{x_{k-1}, \ldots, x_{k-1}}_{m_{k-1} \text{times}}). \tag{2.11b}$$

We still denote the multiplicity of x_i with m_i. The extended partition defines a spline as well, but it will be especially useful to define a basis as seen in the following section. A more advanced and thorough discussion of different aspects of splines can, for example, be found in [80].

2.2.3 B-splines

We now want to introduce a basis for the spline space that is suitable for CAGD.

Definition 2.7. *Let (u_i) be a nondecreasing sequence of real numbers, the* knots. *Let the i-th B-Spline of degree p $B_{i,p}$ be defined as*

$$B_{i,0}(u) = \begin{cases} 1 & \text{if } u_i \leq u < u_{i+1} \\ 0 & \text{otherwise} \end{cases}, \tag{2.12a}$$

$$B_{i,p}(u) = \frac{u - u_i}{u_{i+p} - u_i} B_{i,p-1}(u) + \frac{u_{i+p+1} - u}{u_{i+p+1} - u_{i+1}} B_{i+1,p-1}(u) \tag{2.12b}$$

whereas the quotient $\frac{0}{0}$ is defined to be zero.

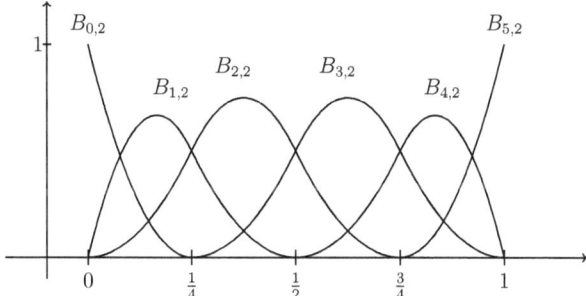

Figure 2.1: B-splines for the knot vector $(0, 0, 0, \frac{1}{4}, \frac{1}{2}, \frac{3}{4}, 1, 1, 1)$

In general the knot sequence is allowed to be bi-infinite, but in the following we will always restrict ourselves to a finite set of knots, the *knot vector* U, and assume that the knot vector forms an extended partition. That especially means that the start and end knot have multiplicity $p + 1$ as in Eq. (2.11a). Instead of denoting the length of the knot vector with $2p + M + 2$ like for the extended partition, we use the length $n + p + 2$ with $n := M + p$, so that we have exactly $n + 1$ B-splines $B_{i,p}$ for $i = 0, \ldots, n$. Figure 2.1 shows some B-splines of degree two. It can be shown that the B-splines are a basis of the spline space defined by the knot vector. Therefore we can alternatively define the space of splines of degree p with knot vector U

$$\mathcal{S}(U) := \operatorname{span}\{B_i\} = \{\sum_{i=0}^{n} B_i \alpha_i : \alpha_i \in \mathbb{R}\}. \tag{2.13}$$

Proofs can e.g. be found in [80] or [77].

An interesting point here is that the smoothness requirement of the spline given by the m_i in Eq. (2.10b) and the multiplicity of values in the extended partition in Eq. (2.11b), respectively, are reflected within the multiplicity of the knots. So at a knot with multiplicity m, while m may range from 1 to p, the spline is $p - m$ times continuously differentiable. Again, for $mp = p + 1$ we allow a discontinuity within the knot. In case of just single knots the whole spline is a C^{p-1} function.

A B-Spline curve is now given by

$$x(u) = \sum_{i=0}^{n} B_{i,p}(u) P_i \tag{2.14}$$

with $P_i \in \mathbb{R}^d$ the $n + 1$ *control points* of the curve.

In the special case where the knot vector only consists of start and end knots, that means

$$U = (\underbrace{u_1, \ldots, u_1}_{p+1 \text{ times}}, \underbrace{u_2, \ldots, u_2}_{p+1 \text{ times}}), \tag{2.15}$$

and $u_1 = 0, u_2 = 1$ the corresponding B-splines degenerate to the Bernstein polynomials,

$$B_{i,p}(u) = b_{i,p}(u). \tag{2.16}$$

The derivative of a B-Spline of degree p is naturally a spline of lower degree $p-1$ and also can be represented by B-splines of degree $p-1$ by the following formula

$$\frac{d}{du}B_{i,p} = \frac{p}{u_{i+p}-u_i}B_{i,p-1}(u) - \frac{p}{u_{i+p+1}-u_{i+1}}B_{i+1,p-1}(u). \qquad (2.17)$$

B-spline Properties

We will now summarize some fundamental properties of B-Splines. Although very basic, this will be very relevant for all the following sections.

- The B-splines for a given knot vector are linearly independent over the interval given by the knot vector.

- B-splines of degree p form a basis for the space of splines of degree p over the same knots.

- B-splines of degree p have the compact support

$$\operatorname{supp}(B_{i,p}(u)) = [u_i, u_{i+p+1}], \quad i = 0, \ldots, n. \qquad (2.18)$$

- B-splines are locally linear independent, e.g. the B-splines that are non-zero on an interval I are linear independent on I.

- B-splines form a partition of unity

$$\sum_{i=0}^{n} B_{i,p}(u) = 1 \qquad \forall u \in [a,b]. \qquad (2.19)$$

- B-splines of degree p are strictly positive over the interior of their support

$$B_{i,p}(u) > 0 \qquad \text{for } u \in (u_i, u_{i+p+1}), \quad i = 0, \ldots, n. \qquad (2.20)$$

The first and last B-spline are equal to one at the start and end knot, respectively

$$B_{0,p}(u_0) = 1, \qquad B_{n,p}(u_{n+p+1}) = 1. \qquad (2.21)$$

- B-spline curves are affine invariant, i.e. the affine map of a B-spline curve is equal to the curve obtained by the affine map of its control points.

Remark 2.8. *The B-spline properties will be revisited from the isogeometric point of view in Sec. 4.2.1.*

Dual Basis

A set of functionals $\phi_j \in \mathcal{S}(\boldsymbol{U})'$ is called *dual basis* if

$$\phi_j B_{i,p} = \delta_{ij}. \tag{2.22}$$

Applied to a spline function $s \in \mathcal{S}(\boldsymbol{U})$ with $s = \sum_{i=0}^{n} B_{i,p} c_i$ this leads to

$$\phi_j s = \phi_j \sum_{i=0}^{n} B_{i,p} c_i = \sum_{i=0}^{n} \delta_{ij} c_i = c_j. \tag{2.23}$$

The dual basis may consists of just function evaluations for the cases with degree 0 or 1, but for general degree it is necessary to use local integrals. For example, we define for a spline s of degree m with knot vector $\boldsymbol{U}$

$$\phi_j s = \int_{u_j}^{u_{j+m}} s(u) \frac{d^m}{dx^m} \Psi_j(x) \, du \tag{2.24}$$

with $\Psi_j(u_j, u_{j+m})$ functions that only depend on two knots. We refer to [80] for the details.

Remark 2.9. *Dual bases are essential for finite elements (see Sec. 3.3.2) and play a role for the convergence analysis in isogeometric analysis in Sec. 4.3.*

Greville Abscissae

For a given knot vector we define the *Greville abscissae* γ_i by averaging the knots

$$\gamma_i = \frac{1}{p} \sum_{k=1}^{p} u_{i+k} \qquad \text{for } i = 0 \ldots n. \tag{2.25}$$

The number of theses coordinates equals the number successive p-tuples of knot intervals and therefore also equal the number of basis functions. Furthermore, they fulfill the property

$$u = \sum_{i=0}^{n} B_{i,p}(u) \gamma_i \text{ for } u \in [a, b]. \tag{2.26}$$

In other words, the Greville abscissa are the control points for representing the identity.

Remark 2.10. *We will make use of the Greville abscissae to represent the basis functions in isogeometric analysis (see Sec. 4.4.1).*

2.2.4 Non-Uniform Rational B-splines

Non-Uniform rational B-Splines (NURBS) are a generalization of the polynomial B-splines to rational splines. With B-splines it is, for example, not possible to represent a circle, which may be represented by trigonometric functions, just like

$$\boldsymbol{C}(u) = (\cos(u), \sin(u)), \quad 0 \leq u \leq 2\pi. \tag{2.27}$$

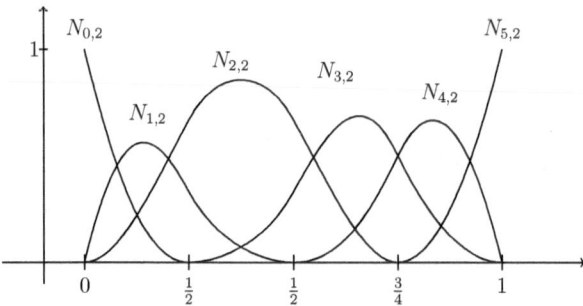

Figure 2.2: NURBS for the knot vector $(0, 0, 0, \frac{1}{4}, \frac{1}{2}, \frac{3}{3}, 1, 1, 1)$ and weights $(1, 1, 1, 2, 1, 1, 1)$

However, the other possibility is using rational functions:

$$\boldsymbol{C}(u) = (\frac{1 - u^2}{1 + u^2}, \frac{2u}{1 + u^2}), \quad 0 \leq u \leq 1 \tag{2.28}$$

The NURBS themselves are defined as

$$N_i(u) = \frac{B_{i,p}(u)\omega_i}{\omega(u)} \tag{2.29}$$

with positive weights $\omega_i \in \mathbb{R}^+$, $i = 0, \ldots, n$ and $\omega(u) := \sum_{j=0}^{n} B_{j,p}(u)\omega_j$. Figure 2.2 shows the example of Fig. 2.1 with changed weights.

NURBS hold very similar properties like B-splines. They are linearly independent, have the same support and form a partition of unity.

The derivatives of NURBS differ from B-Splines, but can be traced back to them. Applying the quotient rule to Eq. (2.29) results in

$$\frac{d}{du}N_{i,p}(u) = \frac{w(u)\frac{d}{du}B_{i,p}(u)w_i + B_{i,p}(u)w_i(\sum_{j=0}^{p}\frac{d}{du}B_{j,p}(u)w_j)}{w^2(u)} \tag{2.30}$$

The connection between B-splines and NURBS lies in homogeneous coordinates $(\omega x, \omega y, \omega z, \omega)$. Given any NURBS curve we can represent it as

$$\boldsymbol{x}^{\omega}(u) = \sum_{i=1}^{n} B_{i,p}(u)\boldsymbol{P}_i^{\omega}. \tag{2.31}$$

This is a B-spline representation in the four dimensional projective space. Therefore, it is possible to generalize some algorithms for B-splines to NURBS. This especially holds for refinement procedures. For more details about NURBS and projective space we refer to [45].

Remark 2.11. *We will meet parameterizations of circles again in Sec. 4.7 and Sec. 5.5. Although NURBS can be interpreted as B-splines in a projective space, there are differences in the evaluation algorithms and not everything implemented for B-splines works in the same manner for NURBS. More details will be given in Sec. 4.6.*

2.3 Multivariate Tensor-product Splines

Until now we have just discussed spline functions in one dimension and have only dealt with curves. To represent surfaces and volumes we construct basis functions in multiple dimensions. A motivation for this is how to obtain a surface out of a curve. We look at a given curve

$$x(u) = \sum_i B_i(u) P_i \qquad (2.32)$$

where the control points themselves are moved along curves, that means

$$P_i(v) = \sum_j B_j(v) P_{ij}. \qquad (2.33)$$

The obtained object is called a swept surface and inserting these equations into each other results in

$$S(u,v) = \sum_{i,j} B_i(u) B_j(v) P_{ij}. \qquad (2.34)$$

Let the B-splines B_{i,p_u} for degree p_u for $i = 0, \ldots, n_u$ defined over the knot vector U with length $n_u + p_u + 2$ and the B-splines B_{j,p_v} of degree p_v for $j = 0, \ldots, n_v$ defined over the knot vector V with length $n_v + p_v + 2$. Then we define the multivariate basis

$$B_{ij}(u,v) := B_i^{p_u}(u) B_j^{p_v}(v), \qquad (2.35)$$

and we obtain the surface representation

$$G(u) = G(u,v) = \sum_i^{n_u} \sum_j^{n_v} B_{ij}(u,v) P_{ij}. \qquad (2.36)$$

Similarly, this can be carried on to higher dimension.

For NURBS the situation is very similar with the help of homogeneous coordinates. If we project them back in Euclidean space, this results in

$$N_{ij}(u,v) = \frac{B_i^{p_u}(u) B_j^{p_v}(v) \omega_{ij}}{\sum_{i,j}^{n_i,n_j} B_i^{p_u}(u) B_j^{p_v}(v) \omega_{ij}} \qquad (2.37)$$

and we get a surface representation

$$G(u) = G(u,v) = \sum_i^{n_i} \sum_j^{n_j} N_{ij}(u,v) P_{ij} \qquad (2.38)$$

The weights ω_{ij} do not follow the tensor-product structure and can be chosen arbitrarily just like the control points P_{ij}. Again this can be analogously formulated in higher dimensions.

The tensor-product basis inherits the properties of the univariate basis, which can easily be shown. So the linear independence, the partition of unity property and the compact support remain unchanged for the multivariate case.

Remark 2.12. *The tensor-product structure will be extensively exploited in the algorithms in Sec. 4.6. On the other hand it is a major obstacle dealing with local refinement (see Chap. 5). The boundary curves and surfaces play a very important role for imposing boundary conditions.*

2.4 Algorithmic Aspects

There are many algorithms dealing with objects based on spline functions. An algorithmically oriented introduction for this topic is [76]. We restrict ourselves only to aspects directly used later on. In Chap. 4 these algorithms will be revisited from the point of view of isogeometric analysis.

2.4.1 Evaluations

The most fundamental algorithm describes how to evaluate a spline object, e.g., a curve or surface. There are several possibilities how to do that:

1. We first evaluate the basis functions and then compute the linear combination with the control points.

2. We directly compute the value with the help of the well known de-Boor-algorithm.

3. We use the control polygon or mesh as an approximation of the object and refine it for a better solution.

For the second and the third approach we refer to [77] since we want to follow the first approach. When looking at the basis function we can make the observation that due to the property of their supports at most $p+1$ functions are nonzero in a knot interval. Precisely, when evaluating them in the interval $[u_i, u_{i+1})$, $i = 0 \ldots n + p$, the basis functions $B_{i-p}, \ldots, B_i$ are nonzero. Using Eq. (2.12b) we can formulate Alg. 2.1. We still assume that $0/0$ is defined to be zero. We point out that this algorithm computes the values of all B-splines efficiently at a single point. Due to the similar structure of the recursion for the B-spline derivatives in Eq. (2.17) the derivatives can also be computed along with the evaluation. We refer to [76] for details of the algorithm. NURBS are evaluated with the help of Eq. (2.29). All necessary B-spline values are provided by Alg. 2.1. The same holds for the NURBS derivatives (see Eq. (2.30)) if the B-spline derivatives are computed beforehand. The evaluation of tensor-product splines can be led back to the evaluation of a univariate spline with the help of Eq. (2.35) or Eq. (2.37). If we want to compute multiple values the algorithm can be constructed to be more efficient than just using Alg. 2.1 several times.

Algorithm 2.1 Evaluation of a B-spline of degree p at value u

Find index k such that $u \in [u_k, u_{k+1})$.
$B_{k,0} = 1$, $B_{j,0} = 0, j \neq k$
for $j = 1, \ldots, p$ **do**
  for $i = k - j, \ldots, k$ **do**

$$B_{i,j} = \frac{u - u_i}{u_{i+j} - u_i} B_{i,j-1} + \frac{u_{i+j+1} - u}{u_{i+j+1} - u_{i+1}} B_{i+1,j-1}$$

  end for
end for
return $B_{k-p,p}, \ldots, B_{k,p}$

2.4.2 Refinement

A spline function is more flexible in regions with more knots than in a region with fewer ones. In order to exploit this freedom it is convenient to be able to refine the spline description of the geometric object. That means that we construct a bigger space that includes the previous space and calculate a new representation of the geometric object without changing it. We have seen in Sec. 2.2.4 that NURBS can be expressed as B-splines in the projective space and therefore all the refinement algorithms are also applicable for NURBS.

As seen in Sec. 2.2 splines are characterized by their knots and their degree. Both aspects can be used for refinement and will be discussed in the following.

Knot Insertion and Knot Refinement

Knot insertion is a basis transformation from the initial basis $B_{i,p}$ defined by the knot vector $\boldsymbol{U} = (u_i)_{i=1,\ldots,n+p+1}$ to the new basis $\widetilde{B}_{i,p}$. This new basis is given by a knot vector

$$\widetilde{\boldsymbol{U}} = (\widetilde{u}_i)_{i=0,\ldots,n+p+2} = (u_0,\ldots,u_k,\widetilde{u},u_{k+1},\ldots,u_{n+p+1}) \tag{2.39}$$

that results from inserting $\widetilde{u} \in [u_k, u_{k+1})$ into $\boldsymbol{U}$. Then it must hold

$$\sum_{i=1}^{n} B_{i,p}(u)\boldsymbol{P}_i = \sum_{i=1}^{\widetilde{n}} \widetilde{B}_{i,p}(u)\widetilde{\boldsymbol{P}}_i. \tag{2.40}$$

By inserting a knot, $p + 1$ of the initial functions are changed. More precisely, this means for $i = 0,\ldots,k-p-1$:

$$B_{i,p}(u) = \widetilde{B}_{i,p}(u). \tag{2.41a}$$

$i = k - p,\ldots,k$:

$$B_{i,p}(u) = \frac{\widetilde{u} - \widetilde{u}_i}{\widetilde{u}_{i+p+1} - \widetilde{u}_i}\widetilde{B}_{i,p}(u) + \frac{\widetilde{u}_{i+p+2} - \widetilde{u}}{\widetilde{u}_{i+p+2} - \widetilde{u}_{i+1}}\widetilde{B}_{i+1,p}(u). \tag{2.41b}$$

$i = k + 1,\ldots,n$:

$$B_{i,p}(u) = \widetilde{B}_{i+1,p}(u). \tag{2.41c}$$

Proofs can be found in [39] or [77].

With some calculations Alg. 2.2 for the computation of the control points $\widetilde{P}_i$ can be derived. Figure 2.3b shows the control points after some knot insertion into the curve in Fig. 2.3a. There are also special algorithms for inserting several knots at once (see [76]).

The extension to multivariate splines is straightforward: Alg. 2.2 is applied in each parameter direction and therefore computes new rows or columns of control points. For efficiency the values α_i may be precomputed.

Algorithm 2.2 Knot insertion algorithm for a B-spline curve of degree p

Find index k such that $\tilde{u} \in [u_k, u_{k+1})$.

for $i = 0, \ldots, n+1$ **do**

$$\alpha_i = \begin{cases} 1 & \text{for } i = 0, \ldots, k-p \\ (\tilde{u} - u_i)/(u_{i+p} - u_i), & \text{for } i = k-p+1, \ldots, k \\ 0 & \text{for } i = k+1, \ldots, n+1 \end{cases}$$

end for

for $i = 0, \ldots, n+1$ **do**

$\quad \tilde{\boldsymbol{P}}_i = \alpha_i \boldsymbol{P}_i + (1 - \alpha_i) \boldsymbol{P}_{i-1}$

end for

return $\tilde{\boldsymbol{P}}_0, \ldots, \tilde{\boldsymbol{P}}_{n+1}$

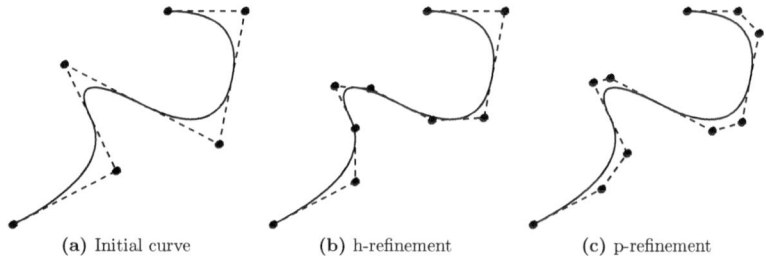

(a) Initial curve   (b) h-refinement   (c) p-refinement

Figure 2.3: Refined curves. The dashed lines are control polygons.

Degree Elevation

Similar to knot insertion degree elevation is a basis change

$$\sum_{i=1}^{n} B_{i,p}(u) \boldsymbol{P}_i = \sum_{i=1}^{\tilde{n}} \tilde{B}_{i,\tilde{p}}(u) \tilde{\boldsymbol{P}}_i \qquad (2.42)$$

from $B_{i,p}$ based on $\boldsymbol{U} = (u_i)_{i=1,\ldots,n+p+1}$ to the new basis $\tilde{B}_{i,\tilde{p}}$ with a degree $\tilde{p} > p$. The knot vector $\tilde{\boldsymbol{U}}$ is created from the knot vector $\boldsymbol{U}$ by increasing every multiplicity m_i, $i = 0, \ldots, n + p + 1$ by one. This ensures that $\tilde{B}_{i,\tilde{p}}$ does not obtain a higher smoothness at these knots so that the initial curve can no longer be represented in this basis. Moreover, all functions are affected and all control points are changed, except those that are interpolatory. There are several algorithmic approaches and different algorithms are possible. We do not go into detail here and refer to [77] or [76]. Figure 2.3c shows the control points after degree elevation of the curve in Fig. 2.3a. Again the application to multivariate splines is done analogously.

Remark 2.13. *Evaluation and refinement will play a crucial role in the implementation of isogeometric analysis. We will use these algorithms in Sec. 4.6 and 5.4.*

Chapter 3

Mathematical Modelling and Finite Element Analysis

This chapter gives a concise introduction to the numerical analysist's point of view of the foundation needed to understand isogeometric analysis. We will begin with mathematical models and the governing partial differential equations. These will be altered to obtain variational formulations, which are fundamental for the numerical solution process either with finite elements or isogeometric analysis. Two sections in this chapter are devoted to the finite element method. We will discuss its theoretical properties as well as its realization and implementation.

Every single topic mentioned above could be treated in detail in its own chapter. So trying to compress it into a single one would be a massive effort and not everything can be elaborated with the extensive depth and therefore the references given have to take over this task. This chapter also fixes the notation for the problem formulation and reviews the finite element features, such that we make a sound comparison with isogeometric analysis. Every aspect of the finite element method that we discuss here will be revisited later. Nevertheless, we will desist from giving much detail about isogeometric analysis, as Chap. 4 is especially devoted to this. Instead we will make some remarks with forward references that do not interfere with the overall exposition and only serve to establish the connection to isogeometric analysis at selected points.

3.1 Mathematical Models

Mathematical models for science and engineering are the foundation for analysis and simulation. The tool to describe physical phenomena are partial differential equations (PDE). Throughout this thesis we restrict ourselves to stationary problems and will only discuss elliptic partial differential equations.

Although tempting we do not want to give too many details in this section and concentrate on the most essential parts, which will also be employed later in the section about numerical analysis. Therefore, we seek suitable formulations for the algorithms right from the beginning.

3.1.1 Heat and Potential Flow

Almost every introduction to partial differential equations starts with the Poisson problem

$$-\Delta u = f. \tag{3.1}$$

It serves as a model equation for different phenomena. Diffusion is described by Fick's first law of diffusion, which relates the diffusive flux f_d with the diffusion coefficient κ and the concentration φ

$$f_d = -\kappa \nabla \varphi. \tag{3.2}$$

If this is combined with the conservation of flux $\nabla \cdot f_d = g$ with a source term g this results in

$$-\kappa \Delta \varphi = g. \tag{3.3}$$

In potential flow we use a potential Φ to describe the velocity $v = \nabla \Phi$. Under the assumption of incompressibility $\nabla \cdot v = 0$ this leads to the Laplace problem

$$\Delta \Phi = 0. \tag{3.4}$$

Remark 3.1. *The Poisson problem will serve as a model problem for local adaptive refinement in Sec. 5.5.*

3.1.2 Continuum Mechanics

Another field we want to investigate are problems from continuum mechanics. It is based upon the assumption that it is feasible to describe a material body as a continuum. This neglects the atomic nature and therefore the body is assumed to be infinitely dividable. We will concentrate on problems from elasticity, which again lead to elliptic partial differential equations.

There are plentiful references that deal with this topic and it is only possible to give an incomplete and subjective selection for continuum mechanics in general [50, 92] or elasticity [72, 31, 75].

Kinematics

At any instant time t a body B occupies an open subset within the euclidean space, a configuration of B. We call the set at the time $t = 0$ the reference configuration Ω_0, which is typically chosen to be the initial state of the body B. A *deformation* $\boldsymbol{\Phi}$ is the mapping from the initial (undeformed) reference configuration Ω_0 to a (deformed) configuration

$$\boldsymbol{\Phi} : \Omega_0 \to \mathbb{R}^3, \tag{3.5}$$

where we assume that $\boldsymbol{\Phi} : \Omega_0 \to \Omega$ is bijective and orientation preserving, that means

$$\det(\nabla\boldsymbol{\Phi}) > 0. \tag{3.6}$$

Each material point can therefore be described by two types of coordinates: the material coordinates in Ω_0 and the spatial coordinates in Ω. The *displacement* $\boldsymbol{u}$ is defined as

$$\boldsymbol{u} = \boldsymbol{\Phi} - \mathrm{id}. \tag{3.7}$$

The relative difference of shapes under deformation is measured by *strain*. One choice is the Green-Lagrange strain tensor, defined as followed

$$\boldsymbol{E} := \frac{1}{2}(\nabla\boldsymbol{u} + \nabla\boldsymbol{u}^T + \nabla\boldsymbol{u}\nabla\boldsymbol{u}^T). \tag{3.8}$$

This expression can also be linearized by neglecting the quadratic terms and we obtain the linearized strain tensor

$$\boldsymbol{\epsilon} := \frac{1}{2}(\nabla\boldsymbol{u} + \nabla\boldsymbol{u}^T). \tag{3.9}$$

Balance Laws

The influence of forces is fundamental to mechanics and is treated via central axioms of conservation. We assume that the configuration Ω is in equilibrium with the applying volume forces $\boldsymbol{f} : \Omega \to \mathbb{R}^3$. Then there exists the traction field $\boldsymbol{t}$ on $S^2 \times \Omega$ so that for each subset $V \subset \Omega$

$$\int_V \boldsymbol{f}(\boldsymbol{x})\, dx + \int_{\partial V} \boldsymbol{t}(\boldsymbol{n}(\boldsymbol{x}), \boldsymbol{x})\, dS = 0, \tag{3.10a}$$

$$\int_V \boldsymbol{x} \times \boldsymbol{f}(\boldsymbol{x})\, dx + \int_{\partial V} \boldsymbol{x} \times \boldsymbol{t}(\boldsymbol{n}(\boldsymbol{x}), \boldsymbol{x})\, dS, \tag{3.10b}$$

where $\boldsymbol{n} : \partial\Omega \to S^2$ denotes the outer unit normal vector to the boundary of Ω. Now it is possible to introduce the stress tensor via following well-known theorem:

Theorem 3.2. (Cauchy's Theorem)
Let $\boldsymbol{t} : S^2 \times \Omega \to \mathbb{R}^3$ *be the traction function for a body* Ω *and suppose* $\boldsymbol{t}(\boldsymbol{n}, \boldsymbol{x})$ *is continuous and that for any sequence of subsets* V_i, *whose volumes tends to zero, it satisfies*

$$\frac{1}{area(\partial V_i)} \int_{\partial B} \boldsymbol{t}(\boldsymbol{n}(\boldsymbol{x}, \boldsymbol{x})\, dS \to 0 \text{ as } vol(V_i) \to 0. \tag{3.11}$$

Then for each $\boldsymbol{x} \in \Omega$ *there is a second order tensor* $\boldsymbol{T}(\boldsymbol{x}) \in \mathcal{V}^2$ *such that*

$$\boldsymbol{t}(n, \boldsymbol{x}) = \boldsymbol{T}(\boldsymbol{x})n. \tag{3.12}$$

$\boldsymbol{T} : \Omega \to \mathbb{R}^3 \times \mathbb{R}^3$ *is called the Cauchy stress tensor field.*

Proof. For example [50], Sec. 3.3.3.   □

Making use of this we can formulate the balance equation (3.10a) with the stress tensor and obtain

$$\int_V \left(\boldsymbol{f}(\boldsymbol{x}) + \operatorname{div} \boldsymbol{T} \right) dx = 0 \tag{3.13}$$

for any $V \subset \Omega$. This results in the differential equation

$$-\operatorname{div} \boldsymbol{T} = \boldsymbol{f}(\boldsymbol{x}). \tag{3.14}$$

The Cauchy stress tensor describes the stress in a deformed configuration. For calculations in structural mechanics it is common to use the Lagrangian description, which refers to the reference configuration. This is done by the first Piola-Kirchhoff stress tensor

$$\boldsymbol{\sigma}^I = \det(\nabla \boldsymbol{\Phi}) \boldsymbol{T} (\nabla \boldsymbol{\Phi})^{-T}, \tag{3.15}$$

which is obtained by a Piola transformation of the Cauchy stress tensor, but is not symmetric. In contrast the second Piola-Kirchhoff stress tensor

$$\boldsymbol{\sigma}^{II} = \det(\nabla \boldsymbol{\Phi}) (\nabla \boldsymbol{\Phi})^{-1} \boldsymbol{T} (\nabla \boldsymbol{\Phi})^{-T} \tag{3.16}$$

is symmetric. More details can be found in [31].

Linear Elasticity

Starting with a linearized theory is not only a common way from the mathematical point of view, but is also a feasible approach if we investigate structures that are only allowed to subject to small deformation and would fail otherwise. In the following we employ the linearized strain tensor ϵ defined in Eq. (3.9). Furthermore, in linearized theory there is no difference between the different stress tensors and therefore we can state the balance law Eq. (3.14) with the second Piola-Kirchhoff stress tensor

$$-\operatorname{div} \boldsymbol{\sigma} = \boldsymbol{f}(\boldsymbol{x}). \tag{3.17}$$

Material Law   A material law connects the strain tensor with the stress tensor. Here we investigate the St.Venant-Kirchhoff material law between the (linearized) Green-Lagrange strain tensor $\boldsymbol{\epsilon}$ and the second Piola-Kirchhoff stress tensor $\boldsymbol{\sigma}$

$$\boldsymbol{\sigma} := \lambda \operatorname{tr}(\boldsymbol{\epsilon}(\boldsymbol{u})) \boldsymbol{I} + 2\mu \boldsymbol{\epsilon}(\boldsymbol{u}) \tag{3.18}$$

with Lamé coefficients λ and μ. These are related to Young's modulus E and Poisson's ratio ν, parameters with more physical meaning, via

$$\mathrm{E} = \frac{\mu(3\lambda + 2\mu)}{\lambda + \mu}, \text{ and } \nu = \frac{\lambda}{2(\lambda + \mu)} \tag{3.19}$$

Sometimes it is more convenient to rearrange the symmetric tensors $\boldsymbol{\epsilon}$ and $\boldsymbol{\sigma}$ into vectors according to the Voigt notation

$$\underline{\boldsymbol{\epsilon}} = \left(\epsilon_{11}, \epsilon_{22}, \epsilon_{33}, 2\epsilon_{12}, 2\epsilon_{13}, 2\epsilon_{23} \right)^T, \tag{3.20}$$

$$\underline{\sigma} = \begin{pmatrix} \sigma_{11}, \sigma_{22}, \sigma_{33}, \sigma_{12}, \sigma_{13}, \sigma_{23} \end{pmatrix}^T. \tag{3.21}$$

Especially the shear components of the strain are scaled by a factor two. The reason for this is to express the tensor product as the scalar product of two vectors as will be used later on. We can now express Eq. (3.18) in matrix notation

$$\underline{\sigma} = C\underline{\epsilon} \tag{3.22}$$

with

$$C = \frac{E}{(1+\nu)(1-2\nu)} \begin{pmatrix} 1-\nu & \nu & \nu & & & \\ \nu & 1-\nu & \nu & & 0 & \\ \nu & \nu & 1-\nu & & & \\ & & & \frac{1-2\nu}{2} & & \\ & 0 & & & \frac{1-2\nu}{2} & \\ & & & & & \frac{1-2\nu}{2} \end{pmatrix}. \tag{3.23}$$

Boundary Conditions In order to complete the formulation, boundary conditions have to be added. For this problem Dirichlet-type fixed displacement boundary conditions

$$\boldsymbol{u} = \boldsymbol{g} \text{ on } \Gamma_D \tag{3.24}$$

or Neumann-type fixed traction boundary conditions

$$\boldsymbol{\sigma n} = \boldsymbol{h} \text{ on } \Gamma_N \tag{3.25}$$

are possible. Furthermore, symmetry boundary condition are a combination of both in the components of the displacement field. For example, to model symmetry to a boundary parallel to the y-axis the x-component has to vanish and the y-component needs to fulfill a Neumann zero condition:

$$u_1 = 0 \tag{3.26a}$$
$$(\boldsymbol{\sigma} \cdot \boldsymbol{n})_2 = 0 \tag{3.26b}$$

For a symmetry boundary parallel to the x-axis it is the other way round.

Lamé Equations By inserting the material law (3.18) in the balance law (3.17) and adding the boundary conditions (3.24) and (3.25) we obtain the strong formulation of the linear elasticity problem

$$-2\mu \text{ div } \boldsymbol{\epsilon}(\boldsymbol{u}) - \lambda \nabla(\nabla \cdot \boldsymbol{u}) = \boldsymbol{f} \text{ in } \Omega, \tag{3.27a}$$
$$\boldsymbol{u} = \boldsymbol{g} \text{ on } \Gamma_D, \tag{3.27b}$$
$$\boldsymbol{\sigma} \cdot \boldsymbol{n} = \boldsymbol{h} \text{ on } \Gamma_N. \tag{3.27c}$$

Other approaches like mixed formulations are not in the scope of this work. We refer to [26] for details.

Plane Stress By applying further assumptions we can simplify the equations and reduce the dimension from three to two. Plane stress implies that we can, without loss of generality, neglect the stress within the third coordinate direction. This is usually the case for an object with uniform low thickness that is very small compared to its other extents. Therefore σ_{33}, σ_{13} and σ_{23} have to be zero. With Eq. (3.22) we see that ϵ_{13} and ϵ_{23} have to vanish as well. As we have assumed that $\sigma_{33} = 0$, we can express ϵ_{33} as followed

$$\epsilon_{33} = \frac{\nu}{\nu - 1}(\epsilon_{11} + \epsilon_{22}) \tag{3.28}$$

By inserting this equation again into Eq. (3.22) the stress-strain relation can be simplified to

$$\begin{pmatrix} \sigma_{11} \\ \sigma_{22} \\ \sigma_{12} \end{pmatrix} = \frac{E}{1 - \nu^2} \begin{pmatrix} 1 & \nu & 0 \\ \nu & 1 & 0 \\ 0 & 0 & \frac{1-\nu}{2} \end{pmatrix} \begin{pmatrix} \epsilon_{11} \\ \epsilon_{22} \\ \epsilon_{12} \end{pmatrix}. \tag{3.29}$$

This model will be used later on in Sec. 4.7.

3.2 Variational Formulation

In order to employ the finite element method it is now necessary to reformulate the problem into a variational or weak formulation. This step will be the same as for isogeometric analysis because it is also based on variational formulations.

The formulations introduced in the previous section lead to classical solutions in the space of continuous functions. Weak formulations are based on weak derivatives and Sobolev spaces. This is not only the appropriate setting for Galerkin projections, but also allows to answer theoretical questions like existence and uniqueness in a coherent way.

We start with an abstract setting and apply this to the problems introduced in the previous section afterwards.

3.2.1 Abstract Setting

Let V be a Hilbert space, and $a : V \times V \to \mathbb{R}$ a bilinear form and $l \in V$. Find $u \in V$ such that

$$a(u, v) = (l, v) \qquad \forall v \in V. \tag{3.30}$$

To ensure existence and uniqueness of a solution certain properties must hold.

Definition 3.3. *A bilinear form $a : V \times V \to \mathbb{R}$ on a normed linear space V is said to be* bounded *or* continuous *if there exists a constant C so that*

$$|a(u, v)| \leq C||u||_V ||v||_V \qquad \forall u, v \in V \tag{3.31}$$

and coercive *on $U \subset V$ or U-elliptic if there exists a coercivity constant $\alpha > 0$ such that*

$$a(u, u) \geq \alpha ||u||_V^2, \qquad u \in U. \tag{3.32}$$

It can be easily seen that a symmetric bilinear form that is continuous and coercive on V induces an inner product and therefore a norm $||v||_E := \sqrt{a(v,v)}$, the energy norm. Furthermore, there is a connection to a minimization problem.

Theorem 3.4. *Let U be a nonempty, closed convex subset of the Hilbert space V, $a : V \times V \to \mathbb{R}$ a symmetric, bounded and V-elliptic bilinear form, $l \in V$ and*

$$E(v) = \frac{1}{2}a(v,v) - (l,v), \quad v \in V. \tag{3.33}$$

Then there exists a unique $u \in U$ such that

$$E(u) = \inf_{v \in U} E(v). \tag{3.34}$$

If $U \subset V$ then u is equivalently defined by

$$a(u,v) = (l,v), \quad \forall v \in V \tag{3.35}$$

Proof. e.g. see [7], Th. 8.3.2   □

In the general case there is no equivalence to a minimization problem but still existence and uniqueness are shown by following Theorem.

Theorem 3.5. *(Lax-Milgram)*
Given a Hilbert space $(V, (\cdot, \cdot))$, a continuous, coercive bilinear form $a(\cdot, \cdot)$ and a continuous linear form $l \in V'$, there exits a unique $u \in V$ such that

$$a(u,v) = l(v) \quad \forall v \in V. \tag{3.36}$$

Proof. e.g. see [27], Sec. 2.7.   □

3.2.2 Application to Models

After this abstract introduction we want to state the variational formulations of the form of Eq. (3.30) for the models introduced in Sec. 3.1. We will also see that the bilinear forms fulfill the requirements of Th. 3.5.

Laplace Equation   By multiplication of Eq. (3.1) with a test function v, integration over the domain Ω and partial integration we get

$$\int_\Omega (\nabla u, \nabla v)\, dx = -\int_\Omega v\Delta u\, dx + \int_\Gamma v(\nabla u, \boldsymbol{n})\, dS. \tag{3.37}$$

For zero Dirichlet boundary conditions only we let $V = H_0^1(\Omega)$ with $|\cdot|_{H^1}$ as the norm (due to the Poincaré-Friedrichs inequality), and define

$$a(u,v) = \int_\Omega (\nabla u, \nabla v)\, d\Omega. \tag{3.38}$$

The continuity and coercivity of the bilinear form are now easily seen as

$$|a(u,v)| \le |u|_{H^1}|v|_{H^1} \tag{3.39}$$

and

$$a(v, v) = |v|_{H_0^1}, \quad \forall v \in V. \tag{3.40}$$

In the case of nonzero Dirichlet boundary conditions $u = g$ on Γ it is assumed that $g \in H^{1/2}(\Gamma)$. This ensures the existence of a trace operator γ with

$$\gamma(H^1(\Omega)) = H^{1/2}(\Gamma) \tag{3.41}$$

and with the help of a function $G \in H^1(\Omega) : \gamma(G) = g$ we can transform the problem into one with zero boundary conditions.

Also pure Neumann boundary conditions $\frac{\partial u}{\partial n} = g$ on Γ_N can be investigated analogously in $H^1(\Omega)$ and mixed boundary conditions within $H^1_{\Gamma_D}$ under suitable assumptions like $l \in L^2(\Omega)$ and $g \in L^2(\Gamma_N)$. More details can be found in [7] or [27].

Linear Elasticity   Starting from the strong form shown in Eq. (3.17) with the stress tensor $\boldsymbol{\sigma}$, we multiply it with a test function $\boldsymbol{v}$ and apply the partial integration rule

$$\int_\Omega \boldsymbol{v} \cdot \operatorname{div} \boldsymbol{\sigma} \, dx = \int_{\partial\Omega} \boldsymbol{v} \cdot (\boldsymbol{\sigma} \cdot \boldsymbol{n}) \, dS - \int_\Omega \boldsymbol{\sigma} : \nabla \boldsymbol{v} \, dx \tag{3.42}$$

with the tensor product $\boldsymbol{A} : \boldsymbol{B} = \operatorname{trace}(\boldsymbol{AB})$. We finally obtain the bilinear form

$$a(\boldsymbol{u}, \boldsymbol{v}) = \int_\Omega \boldsymbol{\sigma}(\boldsymbol{u}) : \boldsymbol{\epsilon}(\boldsymbol{v}) \, dx \tag{3.43}$$

and the linear form

$$l(\boldsymbol{v}) = \int_\Omega \boldsymbol{f} \cdot \boldsymbol{v} \, dx + \int_{\Gamma_N} \boldsymbol{g} \cdot \boldsymbol{v} \, dx \tag{3.44}$$

with help of the identity $\boldsymbol{\sigma} : \nabla \boldsymbol{v} = \boldsymbol{\sigma} : \boldsymbol{\epsilon}(\boldsymbol{v})$, that holds due to the symmetry of the tensors. Here we can make use of the Voigt notation introduced in Eq. (3.20) and (3.21) to write the inner product of the tensors as

$$\boldsymbol{\epsilon}(\boldsymbol{u}) : \boldsymbol{\sigma}(\boldsymbol{v}) = \underline{\boldsymbol{\epsilon}}(\boldsymbol{u}) \cdot \underline{\boldsymbol{\sigma}}(\boldsymbol{v}) = (\boldsymbol{D}_C \boldsymbol{v})^T \boldsymbol{C} \boldsymbol{D}_C \boldsymbol{u} \tag{3.45}$$

with

$$\boldsymbol{D}_C = \begin{pmatrix} \frac{\partial}{\partial x} & 0 & 0 \\ 0 & \frac{\partial}{\partial y} & 0 \\ 0 & 0 & \frac{\partial}{\partial z} \\ \frac{\partial}{\partial y} & \frac{\partial}{\partial x} & 0 \\ \frac{\partial}{\partial z} & 0 & \frac{\partial}{\partial x} \\ 0 & \frac{\partial}{\partial z} & \frac{\partial}{\partial y} \end{pmatrix}. \tag{3.46}$$

Later on we will also use the notation

$$(\boldsymbol{D}_C \boldsymbol{v})^T \boldsymbol{C} \boldsymbol{D}_C \boldsymbol{u} = \boldsymbol{v}(\overleftarrow{\boldsymbol{D}_C} \boldsymbol{C} \overrightarrow{\boldsymbol{D}_C}) \boldsymbol{u} \tag{3.47}$$

where the arrows on top shall indicate whether the operator acts on left or right (see [12]). This will be used again in Sec. 3.5.2.

In this case we can also draw the parallel to the minimization of the energy functional of the system

$$E(\boldsymbol{u}) = \int_\Omega \frac{1}{2} \boldsymbol{\epsilon} : \boldsymbol{\sigma} - \boldsymbol{f} \cdot \boldsymbol{u} \, dx - \int_{\Gamma_N} \boldsymbol{g} \cdot \boldsymbol{u} \, dS \qquad (3.48)$$

as it was formulated in Th. 3.4.

If the stress tensor $\boldsymbol{\sigma}$ is substituted with the material law in Eq. (3.18) or by using the strong form in Eq. (3.27) the bilinear form can alternatively written as

$$a(\boldsymbol{u}, \boldsymbol{v}) = \int_\Omega \lambda(\nabla \cdot \boldsymbol{u})(\nabla \cdot \boldsymbol{v}) + 2\mu\boldsymbol{\epsilon}(\boldsymbol{u}) : \boldsymbol{\epsilon}(\boldsymbol{v}) \, dx. \qquad (3.49)$$

Let $V = [H^1(\Omega)]^d$ and the continuity of the bilinear form can again be shown with the Cauchy-Schwarz inequality. The coercivity can be shown with the help of Korn's inequality

$$||v||^2_{[H^1(\Omega)]^d} \leq c \int_\Omega \boldsymbol{\epsilon}(\boldsymbol{v}) : \boldsymbol{\epsilon}(\boldsymbol{v}) \, dx. \qquad (3.50)$$

Detailed proofs and further information can e.g. be found in [90].

3.3 Finite Element Foundations

The finite element method is one of the predominant and most versatile methods for solving partial differential equations. It is not only capable of solving problems from structural mechanics, although its engineering roots lie there. Numerous monographs are dedicated to different aspects of FEM and we just want to mention [27], [26] and [32] as general comprehensive ones.

The finite element method is based on variational formulations introduced in Sec. 3.2 and then brought into a finite dimensional space by a Galerkin projection. The special framework how to choose the finite dimensional space will be discussed in the following from an abstract perspective.

3.3.1 Galerkin Projection

As described in the previous sections we want to solve following problem: find $u \in V$ so that

$$a(u, v) = (l, v) \qquad \forall v \in V. \qquad (3.51)$$

It is our intent to solve this numerically and therefore we have to use a finite-dimensional space V_h to formulate the finite-dimensional variational formulation: find $u_h \in V_h$, such that

$$a(u_h, v) = (l, v) \qquad \forall v \in V_h. \qquad (3.52)$$

Let $\{\varphi_i\}$ be a basis of V_h then the last formula is equivalent to

$$a(u_h, \varphi_i) = (l, \varphi_i) \qquad \forall \varphi_i. \qquad (3.53)$$

The solution can also be formulated as $u_h = \sum_j q_j \varphi_j$ and this results in the system of linear equations

$$\sum_j a(\varphi_j, \varphi_i) q_j = (l, \varphi_i) \qquad (3.54)$$

which can be written as

$$Aq = b \tag{3.55}$$

with the stiffness matrix A with entries $a_{ij} = a(\varphi_j, \varphi_i)$, the solution vector $q = (q_i)$ and the right hand side $b_i = (l, \varphi_i)$.

In case that the bilinear form $a(\cdot, \cdot)$ induces a norm, we can easily see that the stiffness matrix A is symmetric positive definite, because

$$q^T A q = \sum_{i,j} q_i a_{ij} q_j = a(\sum_j q_j \varphi_j, \sum_i q_i \varphi_i) = \| \sum_i q_i \varphi_i \|_E \geq 0 \tag{3.56}$$

and

$$\| \sum_i q_i \varphi_i \|_E = 0 \Leftrightarrow q = 0 \tag{3.57}$$

due to the linear independence of the basis functions φ_i.

3.3.2 Finite Element Function Spaces

After having introduced the Galerkin projection we now discuss the choice of the basis functions that span the finite-dimensional space V_h. Here we start with the abstract approach similar to [27]. More concrete examples will be given in Sec. 3.4.1.

Definition 3.6. *(Ciarlet)*
Let

- $T \subseteq \mathbb{R}$ *be a bounded closed set with nonempty interior and Lipschitz-continuous boundary* (element domain),

- $\mathcal{P}$ *be a finite-dimensional space of functions on K* (space of shape functions)

- $\Sigma = p_1, p_2, \ldots, p_k$ *be a basis for $\mathcal{P}'$* (the set of nodal variables or degrees of freedom).

Then $(T, \mathcal{P}, \Sigma)$ is called a finite element.

Typically the shape T of the element domain is triangular or quadrilateral (or tetrahedral or hexagonal in three dimensions) and for $\mathcal{P}$ a polynomial space is chosen.

Let φ_i be the basis for $\mathcal{P}$ dual to Σ, i.e. $p_i(\varphi_j) = \delta_{ij}$ (also cf. dual basis for B-splines in Sec. 2.2.3). Furthermore, we call it a *nodal basis* if the functionals are evaluation of any derivative, e.g.

$$p_i(f) := \frac{\partial^{m_i}}{\partial x_{k_i}} f(x_i) \tag{3.58}$$

for fixed $m_i, k_i \in \mathbb{N}$ and $x_i \in \Omega$. For a nodal basis we have

$$\sum_{i=1}^n u_i \varphi_i(x_j) = u_j. \tag{3.59}$$

In other words, the coefficient u_j stands for the numerical solution in x_j and thus carries physical significance. A nodal basis is a typical property of finite elements and some examples are shown in Sec. 3.4.1. This concept of a nodal basis can be generalized to the partition of unity, which is the property

$$\sum_{i=1}^{n} \varphi_i = 1. \tag{3.60}$$

Given a finite element $(T, \mathcal{P}, \Sigma)$, let the set $\{\varphi_i\}$ be the basis dual to Σ. If v is a function for which all $p_i \in \Sigma$ are defined, then we define the *local interpolant* by

$$\mathcal{I}_T v = \sum_{i=1}^{n} N_i(v) \varphi_i. \tag{3.61}$$

After having prepared a single element we will now introduce how elements are put together.

Definition 3.7. *A subdivision $\mathcal{T}$ of a domain Ω is a finite collection of bounded closed sets T_i with nonempty interior and Lipschitz-continuous boundary (i.e. element domains) such that*

- $\operatorname{int} T_i \cap \operatorname{int} T_j = \emptyset \quad \forall i \neq j$

- $\bigcup_i T_i = \overline{\Omega}$

Suppose Ω is a domain with a subdivision $\mathcal{T}$. Assume that $(T, \mathcal{P}, \Sigma)$ forms a finite element and that ℓ is the order of the highest partial derivative involved in the nodal variables. For $f \in C^\ell(\overline{\Omega})$, the *global interpolant* is defined by

$$\mathcal{I}_{\mathcal{T}} f|_{T_i} = \mathcal{I}_{T_i} f. \tag{3.62}$$

We can now define the continuity of the finite element space.

Definition 3.8. *An interpolator $\mathcal{I}$ has continuity order r if $\mathcal{I}f \in C^r(\Omega)$ for all $f \in C^\ell(\overline{\Omega})$ with ℓ the order of the highest partial derivative involved in the nodal variables. We call $V_{\mathcal{I}} := \{\mathcal{I}f : f \in C^\ell(\overline{\Omega})\}$ a C^r finite element space.*

Without further assumptions it is not assured that the global interpolant fulfills any continuity requirements. Moreover, from theoretical and practical points of view a higher regularity of the mesh is necessary.

For the following paragraphs we introduce some notation. For a given element T we define the element size $h_T := \operatorname{diam}(T)$ and for a given subdivision $\mathcal{T}$ we have $h = \max_{T_i \in \mathcal{T}} h_T$. We will also write $\mathcal{T}_h$ to emphasize this relation.

Definition 3.9. *A subdivision $\mathcal{T}$ is called* admissible *if*

- *any edge of any element $T_i \in \mathcal{T}$ is either a subset of the boundary $\partial\Omega$ or also an edge of another element T_j.*

A family of subdivisions $\{\mathcal{T}_h\}$ is said to be regular *if*

- *there exists a κ such that*
$$\kappa \geq h_T/\rho_T \quad \forall T \in \mathcal{T} \,\forall h \tag{3.63}$$
where $\rho_T := \sup\{\text{diam}(S) : S$ is a ball contained in $T\}$

- *the mesh parameter h approaches zero.*

Global continuity is determined by the degrees of freedom at the element boundaries that in turn determine the function on the boundary edges. For an admissible subdivision it is assured that element faces match and coinciding linear forms are identified with the same degrees of freedom. Some examples will be given in Sec. 3.4.1. The regularity constraint on a family of subdivision limits how distorted an element can be and will be used in Sec. 3.3.4.

Finally we can also define the finite element space $S^{p,r}(\Omega, \mathcal{T})$ with degree p and continuity r on Ω for a given admissible triangulation $\mathcal{T}$ by

$$S^{p,r}(\Omega, \mathcal{T}) := \{u \in C^r(\Omega) : u|_{T_i} \text{ is a polynomial of degree } p_i \quad \forall T_i \in \mathcal{T}\}. \tag{3.64}$$

Remark 3.10. *Observe that in FEM the elements and the functions on it are defined locally and define the global approximation space V_h through a subdivision.*

3.3.3 Reference Elements

In the previous section we have seen how different finite elements are arranged in a subdivision and create a global function space. Now, we will review a very important concept for FEM — the reference element. The main idea here is to unify the treatment of the different elements by transforming all elements to a single specific one, which is called *reference element*.

Let $(\hat{T}, \hat{P}, \hat{\Sigma})$ with $\hat{\Sigma} = \{\hat{p}, i = 1 \ldots N\}$ be a finite element and $G : \hat{x} \in \hat{T} \to \mathbb{R}^n$ a uniquely invertible affine mapping. Then we can define another element (T, P, Σ) by

$$\begin{cases} T = G(\hat{T}), \\ P = \{p : K \to \mathbb{R} : p = \hat{p} \circ G^{-1}, \hat{p} \in \hat{P}\}, \\ \Sigma = \{p(G(\hat{a}_i)), i = 1 \ldots N.\} \end{cases} \tag{3.65}$$

We also call these two elements affine equivalent. A family of finite elements is called an *affine family* if all elements are affine equivalent to a single finite element $(\hat{T}, \hat{P}, \hat{\Sigma})$), the *reference element*. In this case it is sufficient to describe the reference element and an uniquely invertible affine mapping G_k for each other element T_k.

The following Lemma studies the relation between functions on two affine equivalent elements. Later on, we will use this for the convergence analysis.

Lemma 3.11. *Let F be a unique invertible affine transformation*

$$F : \hat{T} \to T, \, F\hat{x} = x_0 + B\hat{x} \tag{3.66}$$

with B an invertible matrix and x_0 a vector. For $v \in H^m(T)$ it holds that $\hat{v} := v \circ F \in H^m(\hat{T})$ and

$$|\hat{v}|_{H^m(\hat{T})} \leq C\|B\|^m |\det B|^{-\frac{1}{2}} |v|_{H^m(T)}. \tag{3.67}$$

Proof. Proof of [7] Th. 10.3.3 or [26] Th. 6.6.   □

More details especially about elements that are not affine equivalent can be found in [32]. For general quadrilaterals or even curved boundaries the isoparametric concept, which will be discussed in Sec. 3.4.2.

3.3.4 Convergence Analysis

Solving the variational formulation in a finite dimensional space causes an approximation error. We start with an a priori error estimation where we try to quantify the error with just the information of the weak form and the approximation space, e.g. the finite element spaces introduced before. Following fundamental theorem is the first step into this direction.

Theorem 3.12. *(Céa's inequality)*
Let $a(\cdot, \cdot)$ *be a coercive bilinear form and* u, u_h *solutions of the variational formulations in* V *and* V_h, *respectively. Then it holds*

$$||u - u_h|| \leq \frac{C}{\alpha} \min_{v \in V_h} ||u - v|| \tag{3.68}$$

with the continuity constant C *and the coercivity constant* α *of the bilinear form* $a(\cdot, \cdot)$.

Proof. see [27], Th. 2.8.1.   □

Céa's inequality shows that the error in the Galerkin projection can be estimated by the approximation error within the finite-dimensional space. In order to obtain an error bound for the finite element solution we now review the interpolation error $||u - \mathcal{I}u||$ with the interpolation operator $\mathcal{I}$ introduced in Sec. 3.3.2.

We start by estimating the polynomial interpolation error within the polynomial space

$$P_k := \{p : \boldsymbol{x} \to \sum_{|\alpha| \leq k} c_\alpha \boldsymbol{x}^\alpha\} \tag{3.69}$$

on a single element. This estimate depends on the kind of given reference element $(\hat{T}, \hat{P}, \hat{\Sigma})$.

Lemma 3.13. *(Interpolation Error Estimate on Reference Element)*
Let (T, P, Σ) *be a finite element with* T *star-shaped,* $P_p \subset P$ *and* $\Sigma \subseteq (C^\ell(\overline{T}))'$ *and* $p + 1 - \ell - \frac{d}{2} > 0$ *with dimension* d. *For* $i = 0, \ldots, p + 1$ *there exits a constant* c *such that*

$$|u - \mathcal{I}_T u|_{H^i(T)} \leq c h_T^{p+1-i} |u|_{H^{p+1}(T)} \, \forall u \in H^{p+1}(T). \tag{3.70}$$

Proof. Proof of [27], Theorem 4.4.4.   □

More general propositions about interpolation in Sobolev spaces can be found in [32] or [27]. These kind of estimates belong to the family of the Bramble-Hilbert Lemma.

In order to move from one element to the full subdivision we invoke the reference element concept (introduced in Sec. 3.3.3). We assume that all elements form an affine family with reference element $\hat{T}$ and we can use Lemma 3.11 to show for each single element T that

$$|u - \mathcal{I}_T u|_{H^m(T)} \leq c||\boldsymbol{B}||^{-m}|\det \boldsymbol{B}|^{\frac{1}{2}}|u - \mathcal{I}_T u|_{H^m(\hat{T})}. \tag{3.71}$$

In order to unify the estimate we assume that the subdivision is regular (cf. Def. 3.9). Then it is possible to get upper bounds for the influence of the $\boldsymbol{B}$. This leads to

Corollary 3.14. *(Local Error Estimate)*
For $\mathcal{T}$ be a regular family of finite elements, degree $p \geq 1$ and $0 \leq m \leq p + 1$ it holds

$$||u - \mathcal{I}_T u||_{H^m(T)} \leq ch_T^{p+1-m}|u|_{H^{p+1}(T)}, \quad \forall u \in H^{p+1}(T), \forall T \in \mathcal{T}. \tag{3.72}$$

It is also possible to show this properties under less restrictive regularity assumptions. See e.g. [26] for further details.

The last steps involve extending the estimate to the whole computational domain, which is covered by elements $T_i \in \mathcal{T}$. Therefore we can use the equality

$$||v||_{H^m(\Omega)} = \sqrt{\sum_i ||v||^2_{H^m(T_i)}} \quad \forall v \in H^m(\Omega) \tag{3.73}$$

to show

Theorem 3.15. *(Global Error Estimate)*
For a domain with polygonal boundary, degree $p \geq 1$ and $u \in H^{p+1}(\Omega)$ holds

$$||u - \mathcal{I}u||_{H^m(\Omega)} \leq ch^{p+1-m}|u|_{H^{p+1}(\Omega)}, \quad 0 \leq m \leq p + 1. \tag{3.74}$$

Further general investigations can be found in [27].

Remark 3.16. *The procedure presented here will also serve as a guide for the convergence analysis in isogeometric analysis and it is possible to obtain some similar results, which are presented in Sec. 4.3.*

3.4 Finite Elements in Use

In this section we want to look in particular at more practical aspects in finite element technology. We start by giving some concrete examples for finite elements and discuss their properties and take a look at smooth elements. Then we discuss the isoparametric approach to deal with curved boundaries. The next topics are refinement and error estimation, which are necessary for an efficient simulation.

3.4.1 Some Finite Element Spaces

After having introduced the concept of reference elements we will restrict ourselves to polynomials over reference elements. There is a direct connection between polynomial functions for FEM and interpolation with polynomials introduced in Sec. 2.2.1. If only function evaluations are used to determine the basis functions we speak, just like for

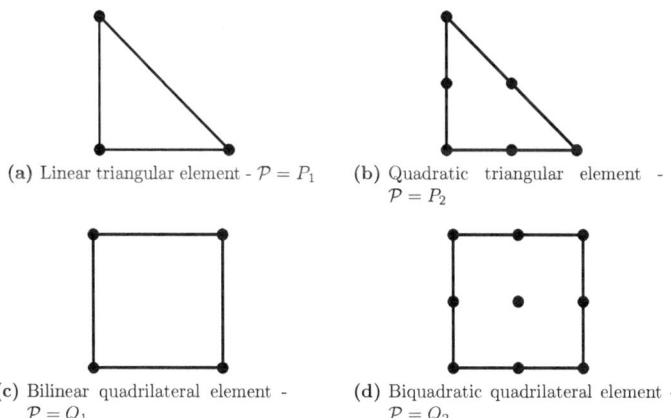

(a) Linear triangular element - $\mathcal{P} = P_1$ (b) Quadratic triangular element - $\mathcal{P} = P_2$

(c) Bilinear quadrilateral element - $\mathcal{P} = Q_1$ (d) Biquadratic quadrilateral element - $\mathcal{P} = Q_2$

Figure 3.1: Lagrange elements - Point values are symbolized by black circles.

Lagrange-Interpolation, of Lagrange elements. If we also make use of derivatives like for Hermite interpolation, we get Hermite elements.

Typically, different polynomial spaces are employed for different types of elements. Triangular elements make use of

$$P_k := \{p : \boldsymbol{x} \to \sum_{|\alpha| \leq k} c_\alpha \boldsymbol{x}^\alpha\} \tag{3.75}$$

whereas quadrilateral elements use

$$Q_k := \{p : \boldsymbol{x} \to \sum_{\substack{\alpha_i \leq k \\ 1 \leq i \leq n}} c_{\alpha_1 \ldots \alpha_n} x_i^{\alpha_1} \ldots x_n^{\alpha_n}\}. \tag{3.76}$$

We will distinguish between C^0 and C^1 elements (see Def. 3.8) in the following.

C^0 Elements

The simplest elements are the Lagrange elements and where $\Sigma = \{p(a_i)\}$ consists only of function evaluations at given coordinates a_i. Fig. 3.1 shows Lagrange elements for triangles and quadrilaterals for degree one and two. All Lagrange elements have C^0 continuity if they are used within an admissible subdivision (see Def. 3.9). Unfortunately just using Hermite elements alone also does not ensure global C^1 continuity and, for example, a simple triangular Hermite element with point and gradient evaluations at the corners is also a C^0 element.

C^1 Elements

The construction of C^1 elements is very involving and we only look at the most simple ones. For triangular elements there is the "Bell triangle" shown in Fig. 3.2a. The function space $\mathcal{P} := \{p \in P_5, \partial_n p \in P_3$ for each side $\}$ is defined by

$$\Sigma = \{\partial^\alpha p(a_i) : |\alpha| \leq 2, i = 1 \ldots 3\}, \tag{3.77}$$

that means the values up to the second derivatives in the corners. A quadrilateral C^1 element is the "Bogner-Fox-Schmit" element shown in Fig. 3.2b. Within this rectangle with the four vertices $a_i, i = 1 \ldots 4$, a polynomial $p \in Q_3$ is uniquely determined by the values, the first derivatives and the mixed second derivatives at these points and therefore

$$\Sigma = \{p(a_i), \partial_1 p(a_i), \partial_2 p(a_i), \partial_{1,2} p(a_i)\}. \tag{3.78}$$

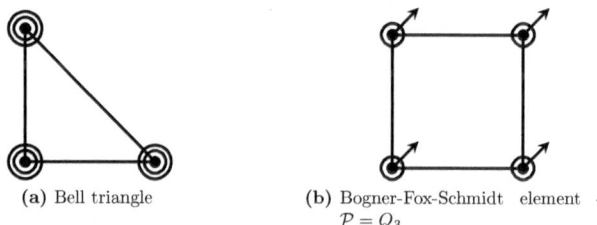

(a) Bell triangle   (b) Bogner-Fox-Schmidt element - $\mathcal{P} = Q_3$

Figure 3.2: Example of C^1 elements: filled black circles symbolized point evaluation and any further circles evaluation of the derivatives, arrows symbolize mixes partial derivatives

On a final note we cite a statement from [101] that emphasizes how complicated it is to create higher continuity within FEM: the lowest degree of a triangular C^m element is equal to $4m + 1$. This especially means that for C^1 continuity we need to use the polynomial degree five just like the Bell element.

Remark 3.17. *In contrast isogeometric analysis offers ansatz spaces of C^1 or even higher smoothness without employing information about the derivatives and with lower degree requirements.*

All elements have in common that the support of each basis function is restricted to elements that contain the corresponding degree of freedom. Examples are shown in Fig. 3.3. Therefore all basis functions have small support and are zero over the remaining domain.

3.4.2 Isoparametric Finite Elements

Approximating a curved boundary of the computational domain is essential and possible when using higher order elements. The isoparametric approach links the basis functions on the element domain with basis functions on the reference element via a polynomial mapping. Let $(\hat{T}, \hat{P}, \hat{\Sigma})$ with $\hat{\Sigma} = \{\hat{p}(a_i), i = 1 \ldots N\}$ and $\boldsymbol{G} : \hat{\boldsymbol{x}} \in \hat{T} \rightarrow (G_j(\hat{\boldsymbol{x}}))_{j \ldots n} \in \mathbb{R}^n$

$$G_j \in \hat{P} \qquad \forall j = 1 \ldots n. \tag{3.79a}$$

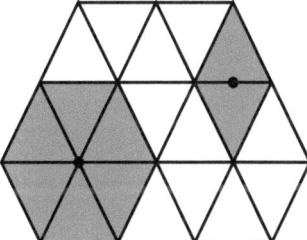

Figure 3.3: Support of finite element basis functions

Then an isoparameteric element (T, P, Σ) is given by (also see Fig. 3.4)

$$\begin{cases} T = \boldsymbol{G}(\hat{T}), \\ P = \{p : T \to \mathbb{R} : p = \hat{p} \circ \boldsymbol{G}^{-1}, \hat{p} \in \hat{P}\}, \\ \Sigma = \{p(\boldsymbol{G}(\hat{a}_i)), i = 1 \dots N\}. \end{cases} \qquad (3.79\text{b})$$

Note that Eq. (3.79a) demands that the mapping is from the same finite element space on the element, which explains the name "iso-parametric". Again we make use of the reference element concept (see Sec. 3.3.3) but here we do not restrict ourselves to affine mappings only. For example we can now use the isoparametric quadrilateral element of degree one that is given by the position of its four corners. With an affine mapping we were only able to create parallelograms.

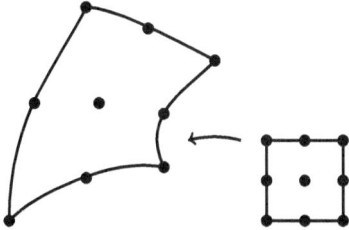

Figure 3.4: Isoparametric finite element

The approximation theory of isoparametric finite element has to take into account the Jacobian of the mapping. In contrast to affine mappings we do not have a simple estimate like Lem. 3.11. Nevertheless, we can interpret the isoparametric element mapping as distortion of the affine element mapping and with some regularity assumptions, like the regularity of the Jacobians, a similar estimate like Th. 3.15 can be derived. For details we refer to [33] and[67].

In practice, isoparametric elements employ quadratic or at most cubic Lagrange-type shape functions, and only edges or faces of elements along a curved boundary are treated this way. Hence interior element boundaries remain flat faces.

Remark 3.18. *Isoparametric elements were created to approximate curved boundaries. This is similar to isogeometric analysis that also takes into account the curved boundaries without approximating them.*

3.4.3 Mesh Refinement

As the statements in Sec. 3.3.4 suggest the quality of the numerical approximation depends on the mesh size h and the ansatz degree p. In order to improve the solution, which was obtained by a computation on a given mesh it is common to increase the number of degrees of freedom by decreasing h or raising p.

Global Refinement

The straightforward approach is to change all elements at once. When looking at a triangle or a rectangular element we can easily refine it by e.g. subdividing it into four pieces — known as red refinement. This is also shown in Fig. 3.5 for a triangular and quadrilateral element. If we apply this to all elements in an admissible subdivision,

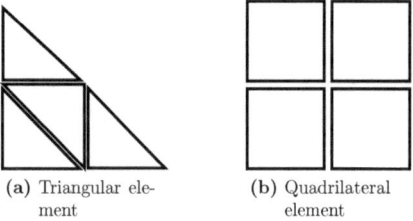

(a) Triangular element   (b) Quadrilateral element

Figure 3.5: Finite element subdivision

we again get an admissible grid with finer resolution. Only if a curved boundary is approximated by the polygonal elements this has to be adapted. This kind of refinement that decreases the mesh size is called *h-refinement*.

Another choice is to increase the ansatz degree and keep the subdivision. Only the basis functions are replaced by ones of higher order, for example when replacing all linear elements(Fig. 3.1a) by quadratic ones (Fig. 3.1b). More details about p-refinement for finite elements can be found in [91]. A combination of h- and p-refinement leads to hp-FEM, see [81].

Local Refinement

If we only want to refine locally because of computational efficiency, this is not straightforward. By refining an element it also happens that it is necessary to refine the neighboring elements as well to get an admissible subdivision. It may also be necessary to use different element shapes as shown in Fig. 3.6a. Another approach is not to demand that the subdivision is admissible and deal with hanging nodes, see Fig. 3.6b. The hanging nodes have to be treated separately and continuity requirements have to be imposed on the system to maintain the continuity. For further discussion we refer to [14] or [15] for the implementation in hp-FEM.

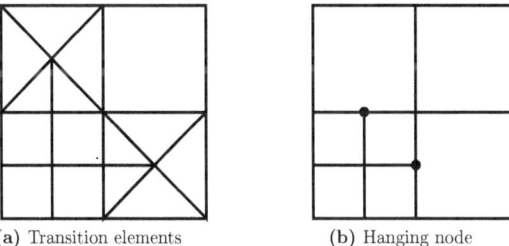

(a) Transition elements   (b) Hanging node

Figure 3.6: Local refinement for quadrilaterals

Remark 3.19. *Although it is not a trivial task, local refinement in FEM is an established techniques and local refinement may be achieved. In isogeometric analysis this is not the case and we will discuss this in Chap. 5 in detail.*

3.4.4 Adaptive Finite Elements and Error Estimation

Until now we have just looked at the refinement procedure, but is also necessary to discuss error estimation, which is an essential component of any adaptive simulation. In Sec. 3.3.4 we have obtained upper bounds for the error of the finite element approximation. Nevertheless, this so-called a priori error analysis is not directly suited to evaluate a numerical computation for several reasons (see [53]). The estimates are valid for sufficient small mesh size h, but it is not known how small it has to be and the constants in the estimate are only known in special cases. Moreover, the assumptions on the function spaces may be too strict and especially the solution may not be smooth enough to belong to $H^{k+1}(\Omega)$. All in all the exact solution is not known and the estimate is not computable.

A Posteriori Error Analysis

To avoid these drawbacks it is necessary to switch to a posteriori error analysis. The idea is to take into account more information and in particular the numerical approximation itself. We want to find a quantitative estimate η for the true error

$$e = u - u_h \tag{3.80}$$

which is unfortunately not available, in a specified norm, with the given numerical approximation u_h and the load and the boundary data. Two aspects here are of special importance: the quantitative one which holds the information how good the numerical approximation is and the spatial one which indicates where regions of lower accuracy are found. Typically the estimate is done element-wise to get the direct connection to the mesh

$$\eta(T) \approx \| (u - u_h)|_T \|. \tag{3.81}$$

Then according to a marking threshold, for example

$$\theta = \alpha \max_k(\eta(T_k)) \tag{3.82}$$

with $\alpha \in [0,1]$, the elements with $\eta(T) \geq \theta$ are marked for refinement. The general adaptive refinement algorithm is shown in Alg. 3.1.

Algorithm 3.1 General adaptive refinement algorithm

compute approximation
estimate error
while tolerance or max. iteration not met **do**
    mark elements according to marking strategy
    adapt mesh
    compute approximation
    estimate error
end while

In order to categorize the estimator two properties are introduced. We call an error estimator η *reliable*, if

$$\|u - u_h\| \leq C\eta. \tag{3.83}$$

That means that the estimator really detects the errors. Nevertheless because this is also the case for a global pessimistic estimate, we call an error estimator *efficient*, if

$$\|u - u_h\| \geq C\eta. \tag{3.84}$$

Both properties are desirable for a posteriori error analysis.

Some Error Estimators

One of the first error estimators introduced by Babuška and Rheinboldt [10] is based on the residual $r(u_h)$ of a PDE and has the form

$$\eta := \left(\sum_{T \in \mathcal{T}} h_T^2 \|r(u_h)\|_{L^2(T)}^2 \right)^{1/2} + \left(\sum_{E \in \mathcal{E}} h_E \| \lceil \frac{\partial u_h}{\partial \nu} \rceil \|_{L^2(E)}^2 \right)^{1/2}. \tag{3.85}$$

Here, h_T is the diameter of the element T, $\mathcal{E}$ the set of all edges, and h_E the length of the edge E. Furthermore, $\lceil \psi \rceil$ is the *jump* of ψ across the edge. Such jumps of the derivative of u_h in normal direction are typical for the standard C^0-continuous basis functions, and they carry significant information with respect to the error.

Averaging error estimates were introduced by Zienkiewicz and Zhou [102]. As seen above, for C^0-continuous basis functions the derivative $p_h := \nabla u_h$ features jumps across the edges between adjacent elements. By averaging a second smoother approximation $\tilde{p}_h \in C^0(\bar{\Omega})$ is constructed and then the difference

$$\eta^A := \|p_h - \tilde{p}_h\|_{L^2(\Omega)^2} \tag{3.86}$$

is employed for error estimation.

The main idea of multilevel error estimation is to enlarge the Galerkin subspace $\mathcal{V}_h$ by another, disjoint subspace $\mathcal{W}_h \subset \mathcal{V}$. This leads to a new subspace

$$\widetilde{\mathcal{V}}_h = \mathcal{V}_h \oplus \mathcal{W}_h \tag{3.87}$$

which is supposed to approximate the solution significantly better. Starting from the residual equation

$$a(e, \psi) = (l, \psi) - a(\varphi_h, \psi) \qquad \text{for all } \psi \in \mathcal{V}, \tag{3.88}$$

we can state a weak formulation for an approximate error $e_h \in \mathcal{W}_h$ (instead of $\widetilde{\mathcal{V}}_h$) : given φ_h, find $e_h \in \mathcal{W}_h$ such that

$$a(e_h, \psi) = (l, \psi) - a(\varphi_h, \psi) \qquad \text{for all } \psi \in \mathcal{W}_h. \tag{3.89}$$

The error estimator η is now defined for each element T via

$$\eta(T) := \| e_h|_T \|_E. \tag{3.90}$$

This estimator is reliable and efficient if the saturation assumption

$$\|u - \widetilde{u}_h\|_E \leq \gamma \|u - u_h\|_E \tag{3.91}$$

with $\gamma < 1$, independent of h, and the strict Cauchy inequality for $a|_{\mathcal{V}_h \oplus \mathcal{W}_h}$ hold with fixed constants in the whole refinement algorithm (for further details see, e.g., [16]). For a simple and efficient implementation, $\mathcal{W}_h$ is chosen as the function space spanned by *bubble functions*. The univariate bubble functions are defined with two real parameters $a < b$ as

$$w[a, b](x) = \begin{cases} \frac{x-a}{b-a} \cdot \frac{b-x}{b-a}, & x \in (a, b) \\ 0, & \text{else} \end{cases} \tag{3.92}$$

Their support is restricted to the interval $[a, b]$, and the extension to two dimensions

$$W_k(x, y) = w[a, b](x) \cdot w[c, d](y) \tag{3.93}$$

has as support the quadrilateral element $[a, b] \times [c, d]$, which simplifies the evaluation of the estimator in Eq. (3.90).

Error estimation for finite elements is a research field on its own and we only want to mention a few references like [4, 94, 30, 11] for further details.

3.5 Implementation Issues

In this section we want to study the implementation aspects of the finite element method. A finite element program has to handle following tasks

1. Obtain a mesh and the corresponding basis function spaces.

2. Assemble the discrete system of equations.

3. Solve the system.

4. Postprocess the numerical solution.

We will deal with the first two problems here. Solving big systems of sparse linear equations is a task on its own within numerical linear algebra and not topic of this thesis. We refer to [48] for iterative solution for large scale problems and to [37] for direct approaches for medium sized sparse systems.

The references about FEM implementations include the engineering point of view [57], MATLAB oriented approaches [5, 6], implementations with focus on triangular grids and adaptivity [79] and quadrilateral grids and adaptivity [13].

3.5.1 Data Representation

In this section we want to discuss the representation of data in finite element analysis. As seen before in Def. 3.7 the computational domain Ω is represented by a grid. This consists of points, which are connected by edges and form elements. Typically it is assumed that Ω has a polygonal boundary, otherwise it has to be approximated.

An example mesh is shown in Fig. 3.7. The atomic components of a mesh are the nodes, which are defined by the coordinates (see Table 3.1a). Elements are stored as an ordered list of nodes as seen in Table 3.1b. This information suffices to define the transformation from a reference element to any element and will be shown exemplary for a linear quadrilateral element later on. Boundary edges are treated separately, because they are needed to enforce boundary conditions. These edges are represented as a list of nodes in Table 3.1c.

In this example the mesh nodes and the degrees of freedom coincide for linear elements. For higher order elements we can introduce further nodes that resemble the degrees of freedom. Especially for isoparametric elements we may use the additional nodes for describing the curved boundary (cf. e.g [17]). It is also possible to separate the definition of the mesh and the degrees of freedom within the element (see [13]). This has the advantage that the information of the mesh can be reused with other types of elements that use other degrees of freedom.

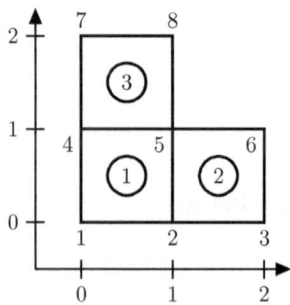

Figure 3.7: FEM mesh example: numbers denote nodes, numbers in circle denote elements

Table 3.1: Data structure for the example FEM mesh in Fig 3.7

(a) Node list		
no.	x-coord.	y-coord
1	0	0
2	1	0
3	2	0
4	0	1
5	1	1
6	2	1
7	0	2
8	1	2

(b) Element list				
no.	nodes			
1	1	2	5	4
2	2	3	6	5
3	4	5	8	7

(c) Boundary list		
no.	node 1	node 2
1	1	2
2	2	3
3	3	6
4	6	5
5	5	8
6	8	7
7	7	4
8	4	1

There are also more advanced data structures that emphasize refinement and also incorporate a tree structure, which holds neighboring information. Details can e.g. be found in [5].

3.5.2 Assembly of System Matrices

The computation of the stiffness matrix and the load vector in Eq. (3.55) — also known as assembly — is an essential algorithmic aspect of FEM. In general it is also the most time consuming part of the simulation program besides the solution of the system of linear equations. For the Poisson equation we can write a matrix entry as

$$a_{ij} = \int_\Omega \nabla\varphi_j \nabla\varphi_i \, dx = \sum_k \int_{T_k} \nabla\varphi_j \nabla\varphi_i \, dx \qquad (3.94)$$

and the load vector as

$$b_j = \int_\Omega f\varphi_j \, dx = \sum_i \int_{T_i} f\varphi_j \, dx. \qquad (3.95)$$

In the following we will omit the load vector, because its assembly is straightforward compared to the assembly of the stiffness matrix and can be done analogously. When evaluating the integrals it would not be wise to integrate over the whole domain because of the small support of each function. Instead it is favorable to assemble the matrices element by element. Furthermore, we still do not need to integrate over all elements, but just over the reference elements and use the transformation rule instead.

Each element has a local numbering of the degrees of freedom defined by the order of the global degrees of freedom in the element list. We denote the mapping from the local indices to the global indices on the element T_k with g_k. An example is given in Fig. 3.8, where we have $g = \{(1, 2), (2, 3), (4, 5), (3.6)\}$.

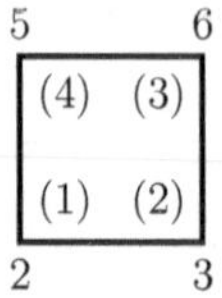

Figure 3.8: Local basis numbers (in brackets) of element No. 2 in Fig. 3.7

Element Stiffness Matrix

Especially, if we look at linear problems and we do not need to take into account any spatial dependencies of coefficients it is possible to further take advantage of the reference element. In this case we are even able to separate the influence of the element mapping and the basis evaluations: the mapped quadrilateral element is given by the coordinates $x_1 \dots x_4$, whereas in case of an affine mapping it is sufficient to use three points to define following constants

$$c_1 = \frac{1}{J}(x_4 - x_1)^2, \tag{3.96a}$$

$$c_2 = -\frac{1}{J}(x_4 - x_1)^T(x_2 - x_1), \tag{3.96b}$$

$$c_3 = \frac{1}{J}(x_2 - x_1)^2, \tag{3.96c}$$

$$J = \det((x_2 - x_1, x_4 - x_1)). \tag{3.96d}$$

On the other hand we can create matrices by (even analytic) evaluation of the bilinear form over the reference element, for example

$$A_1 = \frac{1}{6}\begin{pmatrix} 2 & -2 & -1 & 1 \\ -2 & 2 & 1 & -1 \\ -1 & 1 & 2 & -2 \\ 1 & -1 & -2 & 2 \end{pmatrix}, \tag{3.97a}$$

$$A_2 = \frac{1}{6}\begin{pmatrix} 1 & 0 & -1 & 0 \\ 0 & -1 & 0 & 1 \\ -1 & 0 & 1 & 0 \\ 0 & 1 & 0 & -1 \end{pmatrix}, \tag{3.97b}$$

$$A_3 = \frac{1}{6}\begin{pmatrix} 2 & 1 & -1 & -2 \\ 1 & 2 & -2 & -1 \\ -1 & -2 & 2 & 1 \\ -2 & -1 & 1 & 2 \end{pmatrix} \tag{3.97c}$$

for the Poisson problem and the linear quadrilateral element. Then we can create a the matrix within the range of the local indices, the element stiffness matrix $A^{(T)}$, which is assembled as followed

$$A^{(T)} = c_1 A_1 + c_2 A_2 + c_3 A_3 \tag{3.98}$$

In order to calculate the element stiffness matrix, only the geometry coefficients c_i have to be recalculated. Finally we add each element stiffness matrix to the overall system matrix

$$a_{g(i),g(j)} = a_{g(i),g(j)} + a_{i,j}^{(T)} \qquad (3.99)$$

where g denotes the mapping of local coordinates to global coordinates. For a curved element like in isoparametric FEM, if the mapping is not affine or if some coefficients are spatially not constant, the technique of element stiffness matrices is not applicable.

General Matrix Assembly

If the integrals are not computed in advance or there are spatially dependent, it is necessary to use numerical quadrature rules. Typically we use Gaussian quadrature element-wise with the quadrature points g_i and quadrature weights w_i defined on the reference element

$$\int_T f \, dx = \sum_i f(g_i) w_i \qquad (3.100)$$

Also in the general case, the numerical quadrature for the assembly of the stiffness matrix and the load vector are carried out on one single element $\widehat{T}$. For example, in the case of the Poisson problem this looks like

$$\begin{aligned} a(\varphi_{g(i)}, \varphi_{g(j)}) &= \sum_k \int_{T_k} \nabla \varphi_{g(i)} \cdot \nabla \varphi_{g(j)} \, dx \\ &= \sum_k \int_{\widehat{T}} (DG_k^{-T} \nabla \widehat{\varphi}_i) \cdot (DG_k)^{-T} \nabla \widehat{\varphi}_j) \, |\det DG_k| \, du, \end{aligned} \qquad (3.101)$$

where G_k denotes the transformation from $\widehat{T}$ to T_k, $\widehat{\varphi}_i$ the basis functions on the reference element and g the map of local indices i, j to their global indices on the mesh.

For ease of notation we define the integrand $\alpha(\varphi_i, \varphi_j, T_k)$ so that

$$a(\varphi_i, \varphi_j)|_{T_k} = \int_{\widehat{T}} \alpha(\varphi_i, \varphi_j, T_k) \, du. \qquad (3.102)$$

The arguments φ_i, φ_j symbolize the evaluation of the basis functions (and their derivatives) and the argument T_k symbolizes the transformation of the reference element to T_k. So in the case of Eq. (3.101) we have

$$\alpha(\varphi_i, \varphi_j, T_k) = (DG_k^{-T} \nabla \widehat{\varphi}_i) \cdot (DG_k^{-T} \nabla \widehat{\varphi}_j) \, |\det DG_k|. \qquad (3.103)$$

The basic assembly algorithm is now summarized in Alg. 3.2.

Vector-valued Problems

For vector valued problems like Eq. (3.27) the basis functions φ_i are chosen in a way that only the component $c(i)$ (for a given basis index i) is nonzero:

$$(\varphi_i(x))_\ell = \varphi_i \, \delta_{c(i),\ell} \qquad (3.104)$$

Algorithm 3.2 Basic stiffness matrix assembly algorithm

---

for element index k **do**
  for local basis index i **do**
    for local basis function j **do**
      $a_{\text{loc}} = 0$
      for all quadrature points q in T_k **do**
        $a_{\text{loc}}+ = w_q \alpha(\varphi_i, \varphi_j, T_k))|_q$
      end for
      $a_{g(j),g(i)} = a_{\text{loc}}$
    end for
  end for
end for

Then we can make use of Eq. (3.47)

$$a(\boldsymbol{\varphi}_i, \boldsymbol{\varphi}_j) = \int_\Omega \varphi_{c(i)} \, (\overleftarrow{\boldsymbol{D}} \boldsymbol{C} \overrightarrow{\boldsymbol{D}})_{i,j} \, \varphi_{c(j)} \tag{3.105}$$

and can easily pick the right component of the operator $(\overleftarrow{\boldsymbol{D}} \boldsymbol{C} \overrightarrow{\boldsymbol{D}})$.

3.5.3 Boundary Conditions

Neumann boundary conditions are implemented as additive terms at the load vector. The boundary edges are given by two nodes and the overall integral along Γ_N is subdivided into the sum of line integrals along the element edges $e_i \subset \Gamma_N$

$$\int_{\Gamma_N} g\varphi dx = \sum_i \int_{e_i} g\varphi dx. \tag{3.106}$$

The boundary edges are given by the corresponding corners of the element.

Dirichlet boundary conditions affect the degrees of freedom that are located at the boundary due to the nodal basis property. In order to impose the boundary conditions we can either eliminate the affected degrees of freedom or fulfill the corresponding constraint $\boldsymbol{B}\boldsymbol{q} = \boldsymbol{g}$ by using Lagrangian multipliers

$$\begin{pmatrix} \boldsymbol{A} & \boldsymbol{B}^T \\ \boldsymbol{B} & 0 \end{pmatrix} \begin{pmatrix} \boldsymbol{q} \\ \boldsymbol{\lambda} \end{pmatrix} = \begin{pmatrix} \boldsymbol{f} \\ \boldsymbol{g} \end{pmatrix}. \tag{3.107}$$

Again the corresponding degrees of freedom are stored directly in the boundary list.

3.5.4 Postprocessing

After we having obtained a numerical solution vector $\boldsymbol{q}$ by solving the system in Eq. (3.55) the last step is visualizing the result. Any single function value can be computed by

$$u(\boldsymbol{x}) = \sum_i \varphi_i(\boldsymbol{x}) q_i. \tag{3.108}$$

As introduced in Sec. 3.5 there exists a mesh, which can be equipped with elements that hold the degrees of freedom. These can be used due to the nodal basis property

$$u(\boldsymbol{x}_i) = q_i \qquad (3.109)$$

with $\boldsymbol{x}_i$ positions of the degrees of freedom (see Eq. 3.59), which are also known from the mesh. Further evaluations, if necessary, may be done by interpolation over the element.

Chapter 4

Isogeometric Analysis

It's easier to resist at the beginning
than at the end.

(Leonardo da Vinci)

In this chapter we will present the method called isogeometric analysis based on the previous sections about finite element analysis and computer aided geometric design. Isogeometric analysis, just like FEM, is a method for solving partial differential equations and we use the same models introduced in Sec. 3.1 and the derived variational formulations shown in Sec. 3.2. Especially we can rely on the theoretical properties shown there and the Lax-Milgram theorem still ensures existence and uniqueness of the solution. Also Céa's inequality holds and will be used for an error estimate. Instead of using the FEM function spaces we will rely on the spline spaces introduced for the geometric representation. This tackles the problem of mesh generation and exchanges the triangulations which are adapted to geometric data with the geometric representation itself. As basis functions B-splines or NURBS will be chosen due to their favorable properties.

We will start with the basic idea of isogeometric analysis to give an overview and then we will discuss the method in detail. Due to the fact that we will combine concepts from numerical analysis and FEM with CAGD, new points of view have to be developed that on the one hand rely on established approaches, but on the other hand also resemble the complex relationship between the two origins and the effect of this on the whole method. We will give a short overview of the convergence analysis in isogeometric analysis and discuss the special role of an element in this method. Uniform refinement will be treated separately, as well as some implementation aspects of isogeometric analysis. At the end we will show some numerical examples as well as different applications.

Throughout this section we will often restrict the presentation to two dimensions just out of readability. Typically everything will work similarly in three or more dimension unless stated otherwise.

References for isogeometric analysis in general are not that plentiful compared to the topics of the previous sections. For obvious reason a good starting point is the only monograph [35] currently available. Furthermore, a concise algorithmic introduction can be found in [96]. References for special topics, mainly research articles, will be given at appropriate time.

4.1 Basic Idea and Fundamentals

In the classical FEM the finite dimensional subspace $V_h \subset V$ for the Galerkin projection typically consists of piecewise polynomials defined over a subdivision with global C^0 continuity as it was discussed in Sec. 3.4. Isogeometric analysis is not built on top of this kind of mesh but makes use of the spline space that especially allows higher continuity. Furthermore, the initial geometric description from a CAGD program is already formulated with respect to splines.

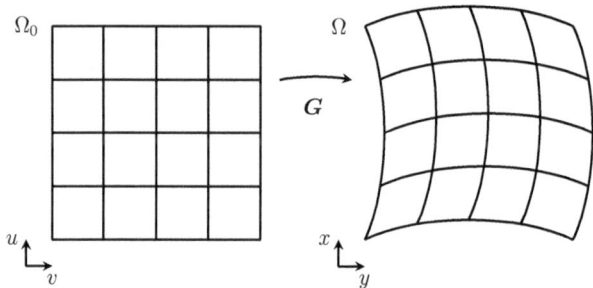

Figure 4.1: Geometry mapping with image of the knot lines

Therefore, our point of departure is a spline parameterization

$$G : \Omega_0 \to \Omega, \quad G(u) = \sum_i N_i(u) P_i \tag{4.1}$$

with control points P_i and with respect to a basis N_i, which maps from the parametric space Ω_0 onto the computational domain Ω (see Fig. 4.1). The basic idea is to formulate the finite dimensional variational formulation

$$a(u_h, v) = (l, v) \qquad \forall v \in V_h \tag{4.2}$$

with respect to basis functions defined on the parameter domain Ω_0 and to use the geometry mapping G from Eq. (4.1) as a global push-forward operator to map these functions to the physical domain Ω. Precisely we will choose B-Spline or NURBS functions N_i, which were introduced in Sec. 2.2.3 and 2.2.4, and were already used to describe the parameterization. Therefore we get the ansatz space

$$V_h = \text{span}\{N_i \circ G^{-1}\}. \tag{4.3}$$

Figure 4.2 visualizes these basis functions on an L-shape domain. The fact that we employ the same basis functions that describe the geometry and are used for the Galerkin projection is the reason why this method was named "iso-geometric" (see [35] for the historical background from the inventors).

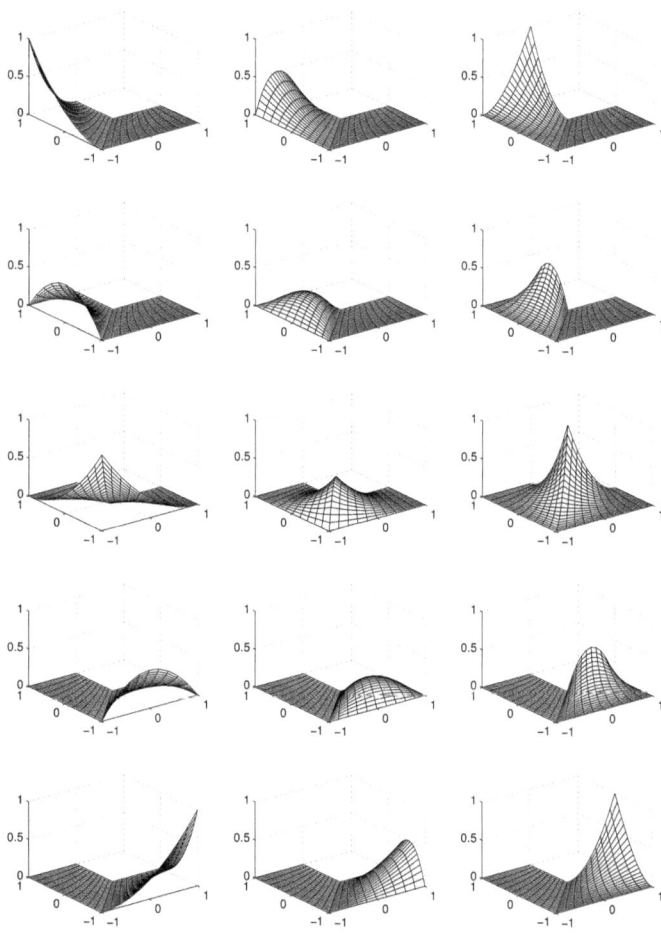

Figure 4.2: Basis functions $N_i \circ \boldsymbol{G}^{-1}$ on an L-shape geometry

4.2 Isogeometric Analysis in Detail

While the basic idea is quite intuitive, we now go further into details of isogeometric analysis and establish the link to previous sections.

Although desirable there is no equivalence to the definition of a finite element (see Def. 3.6) in isogeometric analysis. Still, we can identify following main components

- a set of basis functions N_k,

- an isogeometric mesh $\widehat{\mathcal{T}}$,

- a geometric mapping $\boldsymbol{G} : \Omega_0 \to \Omega$.

Understanding each of these components is fundamental for dealing with advanced isogeometric concepts and we will discuss them in detail in the following.

4.2.1 B-splines and NURBS as Basis Functions

The most obvious difference between isogeometric analysis and FEM is the choice of the basis. Instead of Lagrange or Hermite polynomials we employ B-splines or NURBS. Their properties have been studied in Sec. 2.2 and 2.3. We denote with $\mathcal{B}$ the set of (multivariate) B-splines of degree p_u, p_v for given knot vectors $\boldsymbol{U}, \boldsymbol{V}$ and with $\mathcal{N}$ the set of (multivariate) NURBS for given knot vectors and weights. Then we use

$$\mathcal{S}(\Omega_0) = \mathcal{S}(\Omega_0, p_u, p_v, \boldsymbol{U}, \boldsymbol{V}) := \mathrm{span}(\mathcal{B}((p_u, p_v, \boldsymbol{U}, \boldsymbol{V}))), \tag{4.4a}$$

$$\mathcal{R}(\Omega_0) = \mathcal{R}(\Omega_0, p_u, p_v, \boldsymbol{U}, \boldsymbol{V}, \omega_{ij}) := \mathrm{span}(\mathcal{N}(p_u, p_v, \boldsymbol{U}, \boldsymbol{V}, \omega_{ij})) \tag{4.4b}$$

to denote the corresponding spline spaces.

Due to their essential role in isogeometric analysis it is worthwhile to investigate the effect of B-Splines and NURBS on the Galerkin projection.

- They are *linearly independent*, which ensures that the push forwards of these functions are also linearly independent (under assumption that the geometric mapping is e.g. one-to-one). Furthermore, depending on the bilinear form we may get a symmetric positive definite matrix with the same arguments as in Eq. (3.56) and (3.57). This ensures that the system of linear equations can be solved adequately.

- The choice of spline spaces offers a direct access to *higher order* together with *smoothness* of the basis functions. So it is possible to create an ansatz space with C^1, C^2 or even higher degree, depending on the choice of the knots, without using information about derivatives like e.g. for Hermite elements (cf. Sec. 3.4.1).

- B-splines as well as NURBS fulfill the *partition of unity* property.

- Just like finite element basis functions B-splines or NURBS have a *small compact support* but their support is not only restricted to adjacent elements and its size also depends on the degree.

- Due to the multiplicity of the end knots the univariate splines are interpolatory in these points. If we extend this to a tensor product of functions at least one factor is interpolatory at the boundary. This will be of special importance for boundary conditions. In particular they fulfill the *nodal basis property* $N_i(x_j) = \delta_{ij}$ (see Sec. 2.2.3) at the corners of the parametric domain. Later on, we will make use of this for the implementation in Sec. 4.6.

The influence of the basis functions is not limited to these points and will be discussed further in the next sections.

At this point we make a remark about the indices of the basis functions. We will use to different ways to enumerate tensor product B-splines or NURBS for sake of readability. We will use

- two indices, if we want to emphasize the tensor product structure, e.g.

$$N_{ij}, \ i = 0, \ldots, n_u, j = 0, \ldots, n_v \tag{4.5}$$

- a single index

$$N_k, \ k = 1, \ldots, (n_u + 1)(n_v + 1) \tag{4.6}$$

if we want to emphasize that there is only one set of basis functions. This especially means $N_k = N_{ij}$ for

$$k := 1 + i + j \cdot (n_u + 1). \tag{4.7}$$

It does not play a major role in Galerkin projection from the abstract point of view how these functions relate to each other.

Finally, we want to mention that although we will only restrict ourselves to B-splines and NURBS throughout this thesis, other choices for the basis functions and the spline spaces are possible. For example the usage of generalized B-splines in isogeometric analysis is presented in [71].

4.2.2 Isogeometric Mesh and Relationship to Finite Element Methods

The isogeometric mesh is an integral part of isogeometric analysis which is less visible than the spline basis functions. Nevertheless, it is not of less importance.

We have seen that FEM is based on the description of single elements $(T_i, \mathcal{P}_i, \Sigma_i)$ according to Def. 3.6. Moreover, these elements are typically reduced to reference elements $(\hat{T}, \hat{\mathcal{P}}, \hat{\Sigma})$ and then arranged to construct the finite element spaces shown in Sec. 3.3.2. This can be seen concretely in the techniques for error estimates and in practice by computing element stiffness matrices.

In isogeometric analysis it is just the other way round. We start with a definition of a global space that incorporates the geometry and then reuse B-splines and NURBS as basis functions for the Galerkin projection. Fortunately, although defined globally they also have local support.

Mesh Structure in Isogeometric Analysis

Although it does not originate from a local element definition it is favorable to have a mesh structure in isogeometric analysis. Like for the basis functions we investigate the mesh from the global point of view.

In contrast to FEM the mesh does not have the purpose of defining the geometry (eventually only approximately). Its main task is to subdivide the parametric space Ω_0, where the relation to the computational domain is given by the geometry mapping. The larger inter-element support of the spline basis makes it more involving to find a suitable mesh. If we stick to the way to conclude from the global space to the local properties we can define a subdivision $\mathcal{T}$ that is determined by the intersections of the basis functions.

Definition 4.1. *For a function space over Ω_0 with a basis N_i, we call $\mathcal{T} \subset \mathfrak{P}(\overline{\Omega_0})$ a subdivision with respect to the basis N_i if following conditions are fulfilled.*

- *$\mathcal{T}$ is a subdivision of Ω_0 (see Def. 3.7).*

- *Let $\boldsymbol{u} \in \Omega_0$ and I be an index set, such that $\boldsymbol{u} \in \operatorname{supp} N_i \, \forall k \in I$. In the case that*

$$\operatorname{int}(\bigcap_{k \in I} \operatorname{supp} N_k) \neq \emptyset \tag{4.8}$$

there exists a $T \in \mathcal{T}$ such that

$$T = \bigcap_{k \in I} \operatorname{supp} N_k. \tag{4.9}$$

Equation (4.9) defines each element to be the biggest set within the intersection of the supports. Note that the case of an intersection with empty interior is excluded, because a subdivision does not contain such domains by definition. Definition 4.1 is also valid for finite elements: for any point $\boldsymbol{u}$ within an element domain the basis functions selected in Eq. (4.9) are the basis functions that are defined over the element.

Of course, a pure description only by the supports of the functions is not very usable. Therefore, following lemma shows the connection of the knots within the spline definition.

Lemma 4.2. *For a spline space $\mathcal{S}(\Omega_0, p_u, p_v, \boldsymbol{U}, \boldsymbol{V})$ with a basis $\mathcal{B}$, the subdivision $\mathcal{T}$ with respect to $\mathcal{B}$ consists of Cartesian products of nonempty knot spans.*

Proof. The support of any B-spline $B_{i,j}(u,v) = B_i(u)B_j(v)$ can be expressed as

$$\operatorname{supp} B_{ij} = [u_i, u_{i+p_u}] \times [v_j, v_{j+p_v}], \tag{4.10}$$

for $i = 0 \ldots n_u + p_u, j = 0, \ldots n_v + p_v$ (see Sec. 2.2.3). The intersection of supports that at least have one common point is again a Cartesian product of knot spans $[u_{\hat{i}}, u_{\hat{i}+1}] \times [v_{\hat{j}}, v_{\hat{j}+1}]$. $\qquad\square$

Since B-splines and NURBS share the same support, we can state this Lemma word by word for NURBS. We call this special subdivision $\mathcal{T}$ an *isogeometric subdivision* and each $T \in \mathcal{T}$ an *isogeometric element*.

It should be noted that this is even an admissible subdivision (see Def. 3.9). Nevertheless, the original intention to ensure smoothness does not apply here anymore due to other basis functions and therefore it does not play a role in the isogeometric analysis context.

In certain cases, e.g. Sec. 4.3, it may be useful to only look at open nonempty subsets. Therefore we define the *set of open elements*

$$\mathcal{Q} := \{Q := \operatorname{int} T | T \in \mathcal{T}\}. \tag{4.11}$$

In the following, for the sake of simplicity, we will call any member of $\mathcal{Q}$ or $\mathcal{T}$ *element*. From the context it will always be clear if it is an open or closed element.

Unfortunately, an isogeometric subdivision is not sufficient for a complete description. Due to the use of B-splines we need to take into account multiple knots. Therefore, we call

$$\widehat{T}_{ij} = [u_i, u_{i+1}] \times [v_j, v_{j+1}] \tag{4.12}$$

for $i = 0 \dots n_u + p_u, j = 0, \dots n_v + p_v$ a *knot domain*. Note that we have included empty knot spans and knot spans with empty interior in this definition. These always occur where multiple knots are present and therefore always at the boundary of Ω_0. Precisely, the knot domains $\widehat{T}_{ij}$ for $i \in \{0, \dots, p_u - 1, n_u + 1, \dots, n_u + p_u\}$ and $j \in \{0, \dots, p_v - 1, n_v + 1, \dots, n_v + p_v\}$ always have empty interior and are called *boundary knot domains*. The *isogeometric mesh* is defined as the set of all knot domains,

$$\widehat{\mathcal{T}} := \left\{ \widehat{T}_{i,j} \right\}_{i=0\dots n+p, j=0\dots m+q}. \tag{4.13}$$

Note that this is the equivalent to the extended partition defined for splines (see Sec. 2.2.2) and that this is not a subdivision in the sense of Def. 3.7 due to knot domains with empty interior.

All in all, the elements are the equivalent to elements in FEM and form the support of the basis function. Moreover, they are embedded in the isogeometric mesh which resembles all knots that are used to define and evaluate the basis functions. These mesh concepts will be useful later on and also have their influence on the implementation. In Fig. 4.3 we have visualized the isogeometric subdivision as well as the isogeometric mesh. Since the difference between these meshes lies in knot domains with empty interior, which can not be seen, the empty knot spans have just been drawn close to each other instead of above each other. Finally we can reformulate the support of a basis function with respect to the isogeometric mesh:

Corollary 4.3. *Let* $\mathcal{S}(\Omega_0, p_u, p_v, \boldsymbol{U}, \boldsymbol{V})$ *be a spline space with basis function* $B_{ij} \in \mathcal{B}$. *Then it holds*

$$\operatorname{supp} B_{ij} = \bigcup_{\substack{k=i,\dots,i+p_u \\ \ell=j,\dots,j+p_v}} \widehat{T}_{k,\ell}. \tag{4.14}$$

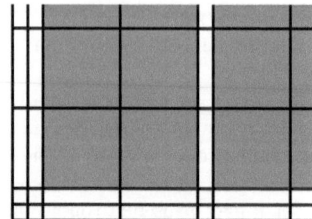

Figure 4.3: Knot domains: Identical knot lines are drawn close to each other. The isogeometric elements are marked in gray and the knot domains with empty interior are marked in white

Multipatch models

Another point of view shall be mentioned here. We do not only look at a single geometry mapping, but allow several of those images, called patches, which form a subdivision of the computational domain. Then we can identify each patch with an element in the sense of Def. 3.6 and the function space is the spline space with its according degrees of freedom. Nevertheless, this tends to hide the difficulties because it does not fit directly into the assembly process described later

Furthermore, if several patches are present they need to be connected similar to finite elements. If we assume that the patches form an admissible subdivision we only have C^0 continuity. Currently it is possible to ensure C^1 differentiability for admissible meshes through constraints on the control points, but this is much more complex for higher continuity. Especially for shells it suffices to obtain a global C^1 surface. More details about this and how different patches are connected in this case are found in [62].

We will give no further details about this viewpoint and just shortly revisit it in the context of local refinement in Chap. 5.

4.2.3 Geometry Mapping – the Relation to CAGD

The geometry mapping G is the component in isogeometric analysis that describes the geometry as it was already introduced in Sec. 2.1.

Role as Reference Domain

By parameterizing the geometry the mapping G establishes the connection between the computational domain Ω and the parametric space Ω_0 where the basis functions are defined. We have already seen this concept, for example, with affine mappings for reference elements in Sec. 3.3.3 or with polynomial mappings for isoparametric finite elements in Sec. 3.4.2.

The element structure described in Sec. 4.2.2 and the spline bases N_k in Sec. 4.2.1 are both defined on the parametric space Ω_0 and are mapped by G to the physical space. Therefore the ansatz space

$$V_h(p_u, p_v, \boldsymbol{U}, \boldsymbol{V}, w_{ij}, \boldsymbol{G}) = V_h(\widehat{\mathcal{T}}, w_{ij}, \boldsymbol{G}) = \mathrm{span}\{N_k \circ \boldsymbol{G}^{-1}\} \qquad (4.15)$$

is influenced by the basis functions as well as the geometry mapping. If we assume
that G is one-to-one, the linear independence of the basis functions is preserved. Also
smoothness is contained if each component of G originates from the same spline space
as the basis functions. Due to the change of coordinates through G the differential
operators have to be transformed as it was already discussed in Sec. 2.1. Using the
transformation rules e.g. the bilinear form of the Poisson problem reads

$$\int_\Omega \nabla \phi_i \cdot \nabla \phi_j \, dx = \int_{\Omega_0} (DG(u)^{-T} \nabla N_i) \cdot (DG(u)^{-T} \nabla N_j \, |\det DG(\eta)| \, du \qquad (4.16)$$

while $\phi_i, \phi_j \in V_h$ with $\phi_i := N_i \circ G^{-1}$ and $\phi_j := N_j \circ G^{-1}$. Although some similarities
with the transformation rule of a finite element in Sec. 3.5, are observable, it should
be stressed that there are two major differences: the integrals refer to single element
domains with simple geometry, and the mapping in these cases is either linear or, in
case of the isoparametric approach, a polynomial. So, we deal with a slightly more
complex setting here.

It comes as no surprise that the geometry mapping also influences the convergence
theory. Under the assumption that the geometry mapping is regular and has a smooth
inverse we can state following estimate

Lemma 4.4. *Let $G : \Omega_0 \to \Omega$ be a diffeomorphism, $Q \in \mathcal{Q}$ and $K = G(Q)$ Then for
all $v \in H^m(K)$, it holds that*

$$|v \circ G|_{H^m(Q)} \leq C || \det \nabla G^{-1} ||_{L^\infty(K)}^{\frac{1}{2}} \sum_{j=0}^{m} ||\nabla G||_{L^\infty(Q)}^j |v|_{H^j(K)} \qquad (4.17)$$

$$|v|_{H^m(K)} \leq C || \det \nabla G^{-1} ||_{L^\infty(K)}^{\frac{1}{2}} ||\nabla G||_{L^\infty(Q)}^{-m} \sum_{j=0}^{m} |v \circ G|_{H^j(Q)} \qquad (4.18)$$

Proof. see [18], Proof of Lemma 3.5   □

Again, this is a similar result to Lemma 3.11 shown before for affine families of finite
elements and we will use it for the convergence theory in Sec. 4.3.

Parametric Violations

There is one issue we want to mention here. Although we always have assumed that
the geometry mapping behaves well, for example, is a diffeomorphism, this might not
always be the case for a parameterization from CAGD. It may even occur that the
parameterization is not even regular in the sense of Def. 2.2. The linear independence
is not influenced if we only have few special singular points, for example, at the boundary. The same problem will occur in Sec. 4.3, where estimates will be shown for a
diffeomorphism G. Numerical experiments seem to indicate that all these assumptions
do not need to be that strict. There are only few references, like [34] or [99], that deal
with the influence of the parameterization. A fully quantitative investigation still lacks.

.

4.2.4 Shared Concepts between the Components

The components introduced are not independent of each other. The strong relationship will be discussed in the following.

Representability of the Geometry

We have seen that the mesh structures are defined to fit to the basis functions just like in Def. 4.1, because we intend to use the element domains to access the basis functions. Furthermore, these basis functions are also not free to choose, because they should always be able to represent the geometry mapping

$$G_i \in \mathcal{S}, \; i = 1, \ldots, d. \qquad (4.19)$$

The reason for this is following. Let $A : \Omega \to \mathbb{R}^d$ be an affine function. Due to the affine invariance of B-splines and NURBS and if Eq. (4.19) holds, $A \circ G$ lies in the spline space and therefore $(A \circ G) \circ G^{-1}$ is an element of V_h. This ensures that the ansatz space V_h contains arbitrary affine functions and therefore the simulation is able to represent them (see e.g. [58]). Furthermore no new representation of the geometry needs to be calculated.

Although the use of NURBS geometries offers more possibilities to represent shapes, this is far more involving in isogeometric analysis than just using B-splines, since the above argument above forces us to use NURBS functions as well in this case. If we use a NURBS geometry in combination with B-splines for the Galerkin projection, we may not be able to represent the geometry any more in the ansatz space.

We can still alter the basis functions and the mesh if we take care that the geometry mapping G is always representable and does not change. This is the initial idea of refinement from the CAGD point of view (cf. Sec. 2.4.2) and we will discuss this again in Sec. 4.5 when dealing with uniform refinement in isogeometric analysis.

Greville Points

If we want to visualize the isogeometric subdivision, we map it to the computational domain via G. In contrast, the basis functions can not be that easily visualized, even in parametric space. One approach makes it comfortable for us to look at the distribution of all basis functions at once: we have seen in Sec. 2.2.3 that Greville abscissae γ_i are the control points for representing the identity for the one-dimensional case. Applied component-wise, we can construct a set of control points with the Greville abscissae in their components. Since the functions are mapped from the parameter space into the physical space by G, we can now map these control points on our computational domain. We call the resulting points *Greville points*

$$\boldsymbol{x}_\gamma = \boldsymbol{G}\big((\gamma_u, \gamma_v)^T\big) \qquad (4.20)$$

The Greville points for an L-shape parameterization are shown in Fig. 4.4. We also use this to visualize the position of NURBS, although we ignore the weights for the calculation of the Greville points. Only the geometry mapping is still a rational spline function.

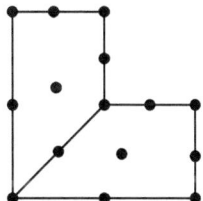

Figure 4.4: Greville points for the L-shape. Basis functions correspond to Fig. 4.2

4.3 Convergence Analysis

Isogeometric analysis and FEM are both based on variational formulations discussed in Sec. 3.2. Therefore Cea's Lemma (Th. 3.12) holds as well and it suffices to estimate the interpolation error. In this section we will obtain an upper bound as it was shown in [18].

It suggests itself to relate the estimates for isogeometric analysis with the ones for finite element analysis, which were presented in Sec. 3.3.4. By showing an estimate for a single element and then relate the elements with each other, the estimate for the interpolation error for the whole subdivision was obtained. Unfortunately, this technique is not directly applicable in isogeometric analysis because the concept of a local element is not present to that extend. Therefore, instead of looking at the interpolation error on an element we have to take into account the whole support of a basis function.

Beforehand, we have to introduce some notation besides the one stated before: for an open isogeometric mesh $\mathcal{Q}_h$ (see Eq. (4.11)) with $h := \max\{h_Q | Q \in \mathcal{Q}\}$ we denote by h_Q the diameter of an element $Q \in \mathcal{Q}$. The image of Q is denoted by $K := G(Q)$ and the set of all elements K in physical space is denoted by $\mathcal{K}$. m_{Q_1,Q_2} denotes the number of continuous derivatives across the common face of two adjacent elements Q_1 and Q_2.

In [18] following space (called "bent Sobolev space") is proposed for the convergence analysis

Definition 4.5. *Let denote* $\mathcal{H}^m$ *the space of functions* $v \in L^2((0,1)^d)$ *within*

$$v|_Q \in H^m(Q), \quad \forall Q \in \mathcal{Q} \tag{4.21a}$$

$$\nabla^k(v|_{Q_1}) = \nabla^k(v|_{Q_2}) \ on \ \partial Q_1 \cap \partial Q_2 \tag{4.21b}$$

for $0 \le k \le \min\{m_{Q_1,Q_2}, m-1\}$ *and for* $Q_1, Q_2 \in \mathcal{Q}$ *with* $\partial Q_1 \cap \partial Q_2 \ne \emptyset$ *and* ∇^k *the* k-th order partial derivative operator.

On $\mathcal{H}^m$ we employ the seminorm

$$|v|_{\mathcal{H}^m} := \sqrt{\sum_{Q \in \mathcal{Q}} |v|^2_{H^i(Q)}}, \quad 0 \le i \le m \tag{4.22}$$

and norm

$$\|v\|_{\mathcal{H}^m} := \sum_{i=0}^{m} |v|_{\mathcal{H}^i}^2 \qquad (4.23)$$

where $\tilde{Q}$ denotes union of the supports of all basis functions that intersects Q, the so called *the support extension* of Q. In the same way, we denote the image of the support extension with $\tilde{K} := \boldsymbol{G}(\tilde{Q})$.

First of all, it is necessary to obtain an interpolation estimate for NURBS in parametric space. Therefore a projector $\mathcal{I}_{\mathcal{S}_h} : L^2((0,1)^d) \to \mathcal{S}_h$ is defined with the help of dual basis (see Sec. 2.2.3) for B-splines and extended to NURBS (see [18] and the references therein for details). For this projector $\mathcal{I}_{\mathcal{R}_h} : L^2((0,1)^d) \to \mathcal{R}_h$ we get the estimate

Lemma 4.6. (Error Estimate in Parametric Space)
Let m and l be integers indices with $0 \le m \le \ell \le p+1$. Then it holds

$$|u - \mathcal{I}_{\mathcal{R}_h} u|_{H^m(Q)} \le c h_Q^{\ell-m} \|u\|_{\mathcal{H}_h^\ell(\tilde{Q})} \qquad \forall u \in \mathcal{H}_h^\ell, \forall Q \in \mathcal{Q}_h \qquad (4.24)$$

with a constant c that depends only on the NURBS weights.

Proof. cf. [18], Proof of Lemma 3.4. □

This is similar to the interpolation error on reference elements for FEM (see Lem. 3.13). We now define the projector into $V_h = \operatorname{span}\{N_i \circ \boldsymbol{G}^{-1}\}$ via

$$\mathcal{I}_{V_h} : L^2(\Omega) \to V : \mathcal{I}_{V_h} v := (\mathcal{I}_{\mathcal{R}_h}(v \circ \boldsymbol{G})) \circ \boldsymbol{G}^{-1}. \qquad (4.25)$$

With the help of the estimate in parametric space in Lem. 4.6 and the estimate for the influence of the geometry mapping in Lem. 4.4 we obtain

Theorem 4.7. (Local Error Estimate)
Let m and ℓ be integer indices with $0 \le m \le \ell \le p+1$, $Q \in \mathcal{Q}_h$, $K = \boldsymbol{G}(Q)$. Then it holds that

$$|u - \mathcal{I}_{V_h} u|_{H^m(K)} \le c h_K^{\ell-m} \sum_{i=0}^{\ell} \|\nabla \boldsymbol{G}\|_{L^\infty(\tilde{Q})}^{i-\ell} |v|_{H^i(\tilde{K})}, \qquad \forall u \in L^2(\Omega) \cap H^\ell(\tilde{K}) \quad (4.26)$$

with $h_K = \|\nabla \boldsymbol{G}\|_{L^\infty(Q)} h_Q$.

Proof. cf. [18], Proof of Th. 3.1. □

Just like for Th. 3.15 we sum up the local error estimate in Th. 4.7 that leads to

Theorem 4.8. (Global Error Estimate)
Let m and ℓ be integer indices with $0 \le m \le \ell \le p+1$. Then it holds that

$$\sum_{K \in \mathcal{K}_h} |u - \mathcal{I}_{V_h} u|_{H^m(K)}^2 \le c \sum_{K \in \mathcal{K}_h} h_K^{2(\ell-m)} \sum_{i=0}^{\ell} \|\nabla \boldsymbol{G}\|_{L^\infty(\boldsymbol{G}^{-1}(K))}^{2(i-\ell)} |u|_{H^i(K)}^2 \quad \forall u \in H^\ell(\Omega).$$

$$(4.27)$$

Compared to the global error estimate for FEM in Th. 3.15 the same optimal rate of convergence is obtained for $\ell = p+1$. Nevertheless the geometry mapping $\boldsymbol{G}$ has a much higher influence here.

These estimates depend on the mesh size h. Further estimates for increasing the degree p with higher smoothness are more advanced and can be found in [22].

4.4 Isogeometric Element Concept

Until now it was sufficient to look at the isogeometric element as a subset of the parametric domain and the set of all elements, the isogeometric subdivision. This is the global point of view. In this section we want to investigate the local point of view and start looking at one single element and its connections. As discussed in Sec. 4.2.2 the basis functions affect more elements than in FEM in exchange for smoother basis functions. Furthermore, it is not possible to define an isogeometric element fully separate from the other ones. Nevertheless, there are some common properties we want to focus on and construct an isogeometric reference element.

4.4.1 Isogeometric Reference Element

The role of the reference element in FEM as seen in Sec. 3.3.3 is a strong one. The treatment of a family of elements that are related through transformations is reduced to a single one. Modulo this transformation they all share the same domain or shape, the same functions and the same degrees of freedom.

In isogeometric analysis the situation is more complex. In Sec. 4.2.2 we have introduced the concept of an isogeometric subdivision, which consists of elements, and the isogeometric mesh, which consists of knot domains. The isogeometric subdivision was defined to consist of sets that do not have empty interior. These are not suitable to be the domain for functions. Unfortunately, it is not possible to describe the basis functions only with the isogeometric subdivision and therefore the isogeometric mesh comes into play.

Single Reference Element

We will now inspect the situation in isogeometric analysis in detail for a single element. In Sec. 4.2.2 we have defined the elements $T \in \mathcal{T}$ in the parametric space. Although they are all rectangular the functions upon them may be very different. Even if we look at an uniform knot vector with only single knots the basis functions differ as shown in Fig. 4.5 for a spline of degree two. In particular this is the case for the boundary elements due to the multiple start and end knots. In Fig. 4.5a we see a boundary element, whereas Fig 4.5b shows the element to its right. Even for this simple one-dimensional case we can easily observe that the basis functions are not the same.

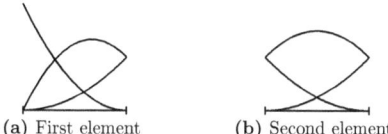

(a) First element   (b) Second element

Figure 4.5: Differences in isogeometric elements

So at first glance it may seem that we cannot find a common ground between the elements in isogeometric analysis. Although we do not have strong properties as before, the isogeometric reference element illustrates the connection between the element and its basis functions. The distribution of the basis functions over the mesh seem arbitrary (especially compared to the very restrictive finite elements), but there is a structure we want to investigate.

Lemma 4.9. *Let $\mathcal{T}$ be a isogeometric subdivision. There are exactly $(p_u + 1)(p_v + 1)$ nonzero basis function within each element $T \in \mathcal{T}$.*

Proof. The support of any basis function consists out of $(p_u + 1)(p_v + 1)$ knot domains according to Cor. 4.3. Therefore any knot domain that is not a boundary knot domain is in the support of $(p_u+1)(p_v+1)$ basis functions. Hence, this holds for all elements. □

Reference Element in the Mesh

The isogeometric element alone is not that informative as for example in FEM. It is more important to look at the whole patch of elements, but even then the distribution of the degrees of freedom may be confusing.

In Fig. 4.6a the support of an univariate B-spline of degree two is shown in grey. It covers three knot domains, which may also be empty although they are not drawn like this. The B-spline itself is marked by a white dot. In Fig. 4.6b all basis functions that are non-zero over an element (marked black) are drawn. We see that the element assumes different positions within each basis function's support: for the leftmost basis function the element is its rightmost support knot domain. For the basis function at the right, it is the other way round and for the one in the middle the element is in the middle position of the support. This results in the reference element are shown in Fig. 4.6c. For any element we have three degrees of freedom that are ordered from left to right.

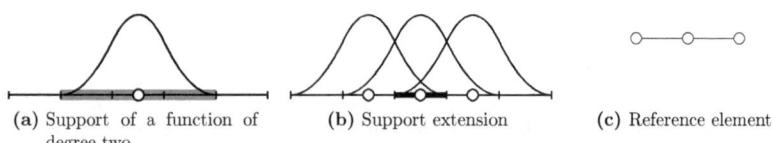

(a) Support of a function of degree two   (b) Support extension   (c) Reference element

Figure 4.6: Univariate B-splines over knot domains

A similar situation is shown for two dimensions in Fig. 4.7 for $p_u = p_v = 2$. According to Cor. 4.3 the support spreads over nine knot domains as visualized in Fig. 4.7a. Figure 4.7b shows the support extension of the element in the middle and all basis functions that are non-zero on it. We see again that the distribution of the basis functions in relation to the element position forms a regular grid of 3×3 positions. The reference element for the two-dimensional case with $p_u = p_v = 2$ is shown in Fig. 4.7c where we have a ordered grid of 3×3 degrees of freedom. We can conclude that not only the number of nonzero basis function over an element is constant as shown in Lem. 4.9, but also how these function are placed related to the element. Just

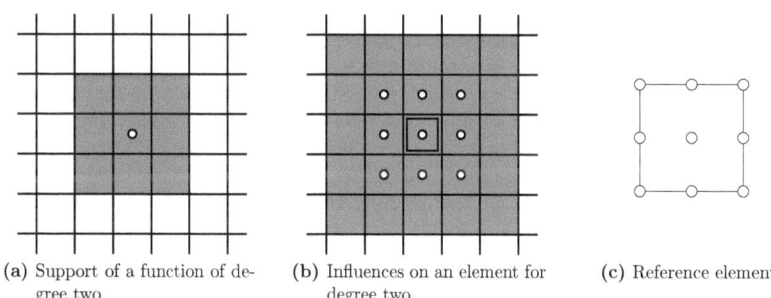

(a) Support of a function of degree two

(b) Influences on an element for degree two

(c) Reference element

Figure 4.7: Bivariate B-splines over knot domains

like the element T takes a unique position in the support of a basis function that is non-zero over T, each basis function takes a position with respect to the element.

Corollary 4.10. *Let $\mathcal{T}$ be a isogeometric subdivision and $\widehat{\mathcal{T}}$ the isogeometric mesh. For any element T the nonzero basis functions over T are B_{ij} with $i = k - p_u, \ldots, k, j = \ell - p_v, \ldots, \ell$, whereas k and ℓ are chosen so that $T = \widehat{T}_{k,\ell}$.*

Proof. Follows with the arguments in the proof of Lem. 4.9 directly out of Cor. 4.3. □

Note that this is valid for any element independent of its exact shape or position. We want to stress here that this also shows the usefulness of the isogeometric elements and the knot domains. On elements we can make use of the information about the basis functions that are non-zero but the support of these basis functions spreads over knot domains. The difference between the two cases in Fig. 4.5 was caused by empty knot spans. This changes the shape of the basis functions but not their support with respect to knot domains. Therefore we can identify the abstract structure within an element independently of its neighboring elements or knot domains.

Although this looks similar to, for example, the biquadratric quadrilateral element (see Fig. 3.1d) it is important to keep in mind that the shape and the support of the basis functions are different. Especially there are no boundary degrees of freedom, because all degrees of freedom in isogeometric analysis affect the neighboring elements.

It might seem that the concept of the isogeometric reference element is not of wide influence because there is no information what the local basis functions values are and how they may be computed. However, this is not in the scope of a single element but a task of the global spline space or at least the support extension of the element. The abstract information within the isogeometric reference element suffices to build the implementation very similar to FEM, as it will be shown in Sec. 4.6. Furthermore it will be of particular importance in Sec. 5.3.3 for local refinement.

4.4.2 Bézier Extraction

Bézier extraction [25] is another element concept and mainly used to bridge the gap between the isogeometric approach and existing FEM codes. As discussed before one therefore has to choose a representation of the basis functions that is more local than B-splines. The Bernstein polynomial introduced in Sec. 2.2.1 are a suitable choice for this. For all B-splines over an element the representation in Bernstein polynomials is computed by knot insertion. Nevertheless, the transformation does not eliminate the complexity of the problem. The locality of the basis is exchanged for additional constraints that enforce continuity and smoothness. Furthermore, additional effort for NURBS has to be made. Due to the fact that it is not our first interest to connect isogeometric analysis with FEM methods on an implementation level, we will not go further into detail regarding this topic.

4.5 Uniform Refinement

Refinement for an object parameterized by spline functions was already introduced in Sec. 2.4.2. Here we restrict the discussion to uniform refinement because we only need to reinterpret the concepts from CAGD in isogeometric analysis. This is not the case in local refinement where even more sophisticated methods are needed. Chapter 5 is devoted to this problem. Typically uniform refinement is applied before the simulation starts, because in most cases the geometric parameterization is too coarse for the use in numerical simulation. Compared to the whole simulation process refinement is executed fast.

Before starting with the details we want to give an important remark about the different viewpoints of refinement. The main focus of Sec. 2.4.2 was to find a new representation of a given spline object (represented by its control points). In context of isogeometric analysis or finite elements we are mainly interested in increasing the approximation space and therefore get more basis functions and degrees of freedom. From this point of view calculating the representation of the geometric object in the new basis is of secondary interest. It suffices that the geometric mapping always remains unchanged out of reasons already stated in Sec. 4.2.4.

It is much more important to create additional degrees of freedom. How this is done depends on the kind of refinement, but it is necessary to study the effect on the overall space, because the numerical solution is determined within it.

h-Refinement   As discussed earlier h-refinement is the process of reducing the mesh width in classical finite elements (see Sec. 3.4.3). The equivalent in isogeometric analysis is knot insertion as it was described from the CAGD point of view in Sec. 2.4.2. So by adding knots within each element we subdivide it into smaller elements, just like in FEM. From the point of view of splines knot insertion reduces the continuity requirements at the insertion point and the overall spline space gets bigger. Figure 4.8 shows the h-refinement applied to the L-shape geometry by inserting a knot in the middle of each nonempty knot span.

p-Refinement   Increasing the degree of the ansatz function, also called p-refinement (see Sec. 3.4.3), is the equivalent to degree elevation. Just like in FEM the element domains are not altered during p-refinement, but in contrast the hierarchical mesh is changed. As it was described in Sec. 2.4.2 the multiplicity of each knot is increased

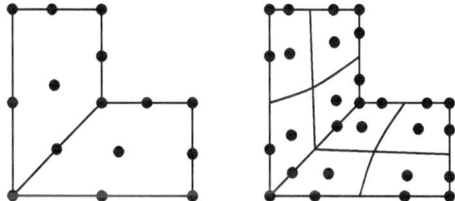

Figure 4.8: h-refinement: initial parameterization with Greville points (left), h-refined parameterization (right)

by one. This is important because an increase of smoothness has to be prevented to ensure that the old function space is still included in the new space. Interpreted from the view of the isogeometric mesh an empty knot domain is inserted between two elements and at the boundaries. Similar to a finite element the degrees of freedom within each isogeometric element increase. Figure 4.9 shows the p-refinement applied to the L-shape parameterization, which raises the degree from two to three.

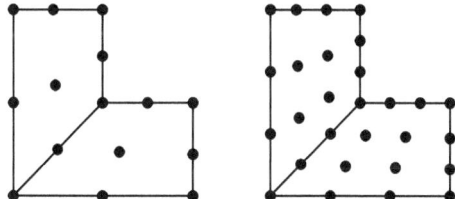

Figure 4.9: p-refinement: initial parameterization with Greville points (left), p-refined parameterization (right)

k-Refinement   Introduced in [58] k-refinement is a combination of p- and h-refinement. Due to the higher inter-element continuity the order of application of p- and h-"re"-fine"-ments are applied makes a difference. We will only use k-refinement for creating a suitable start configuration from initial geometry mapping. In order to maintain the smoothness p-refinement is always applied before h-refinement. This also has the effect that the smoothness in the newly added knots is higher when we have applied p-refinement before. Further background to k-refinement as a refinement method can be found in [44].

4.6 Implementation Issues

In this section we want to investigate some algorithmic aspects in isogeometric analysis based on the implementation issues already discussed separately for CAGD in Sec. 2.4 and FEM in Sec. 3.5. Although the realization of isogeometric analysis shares a lot with its parent fields there are more algorithmic and in-depth issues that are worth looking at. The intention of this section is twofold: On the one hand it presents the main ingredients to a concrete realization of an isogeometric solver and on the other hand it illustrates the differences between the algorithms in CAGD and in numerical analysis.

4.6.1 Data Representation

In FEM we represent the computational domain by a mesh. On the mesh itself elements with different degrees of freedom are defined. In isogeometric analysis this is different: the main information is given by a parameterization and therefore the data structures are the following: A parameterization is given by

- knot vectors - a list of values,

- control points -a list of vectors,

- weights - a list of positive values.

An example is shown in Tab. 4.1 for the L-shape. The control points can be arranged into a control grid. Although the control grid is necessary to describe the geometry it should be noted that it does not serve any further purpose. The isogeometric mesh is given by the knot vectors, which define the knot domains. Also the isogeometric elements do not need to be explicitly specified, because they are simply the knot domain with nonempty interior.

Table 4.1: Data structures for the parameterization of an L-shape

(a) Knot vectors

U	0	0	0	0.5	0.5	1	1	1
V	0	0	0	1	1	1		

(b) Control points and weights

C_1	-1	-1	-1	0	1	-0.6	-0.55	-0.5	0	1	0	0	0	0.5	1
C_2	1	0	-1	-1	-1	1	0	-0.5	-0.55	-0.6	1	0.5	0	0	0
ω	1	1	1	1	1	1	1	1	1	1	1	1	1	1	1

Just like in finite elements we have to associate the boundary of the domain with boundary conditions. Instead of boundary edges we have boundary curves that are obtained by restricting the geometry mapping to the boundary of the parametric domain. Compared to the finite element data structure we do not have any irregular information here and the basis functions or the degrees of freedom are only given indirectly.

4.6.2 Element Structure and Enumeration

Before we go into detail with the algorithms, we have to discuss how the data for isogeometric analysis is structured. In contrast to FEM we deal with regular structures like the tensor product on elements as well as function level. As a side effect, there is no direct enumeration given by the data for isogeometric analysis. So there is a lot of implicit ordering within the data, which glues together the mesh structure with the basis function defined upon it.

Isogeometric Mesh

Enumerating the elements is different than for finite elements. We do not want to take into account all knot domains that may have empty interior and restrict ourselves only to isogeometric elements. On the other hand the position of the knot domains is still necessary for evaluating the basis functions. In the sense of the isogeometric mesh, the knot vectors U, V define the knot domains. The isogeometric subdivision is obtained by removing multiple knots. In order to still have the information of the isogeometric mesh the knot domain index for each element is saved. So for each $T \in \mathcal{T}$ we store i, j such that $T = \widehat{T}_{ij}$ with $\widehat{T}_{ij} \in \widehat{\mathcal{T}}$.

An example is shown in Fig. 4.10, where the isogeometric mesh for the data from Tab. 4.1a is shown. We see $9 \cdot 7 = 63$ knot domains and thereof two elements with indices $(3, 3)$ and $(5, 3)$.

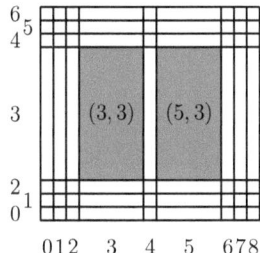

Figure 4.10: Knot span enumeration for the L-shape parameterization

Enumerating the Tensor Product Structure

We have seen that the tensor product structure can be found in different scopes:

- basis functions N_{ij} (see Sec. 4.2.1),

- isogeometric elements $T \in \mathcal{T}$ (see Sec. 4.2.2),

- local basis functions within an element (see Sec. 4.4.1).

For sake of brevity and because they will be used in loops later on we enumerate these entities with a single number. A straightforward way to number a tensor-product structure with n_u entities in u-direction is

$$f(i_u, i_v) := i_u + (i_v - 1)n_u \qquad (4.28)$$

whereas $i_u = 1, \ldots, n_u$ describes the coordinate number in u-direction and $i_v = 1, \ldots, n_v$ in v-direction. Of course this is not the only possibility but we will stick to this to demonstrate how we can make use of the structure within isogeometric analysis. It should be noted that if it is useful to change the enumeration of the degrees of freedom to influence the matrix structure or the element ordering for technical reasons like cache-memory access it is still possible to add these permutations in between. It is important that although we introduced this sequential enumeration we can still access everything by coordinates.

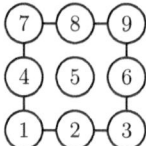

Figure 4.11: Local enumeration of the basis functions for an element of degree $(p_u, p_v) = (2, 2)$

For each element we have a fixed number of basis functions, to whom we assign a local enumeration as shown in Fig. 4.11. For each local basis number we need to compute the global basis number. This is done with the help of the knot span indices and Cor. 4.10 that describes the structure of the non-zero basis functions.

4.6.3 Matrix Assembly

For the assembly of the stiffness matrix and the load vector it is very helpful to have the information provided by a mesh. This allows us to proceed similar to the assembly for FEM in Sec. 3.5.2.

Basic Procedure

Although the integrals are defined over the whole domain, it would be inefficient to compute them this way. Instead we use an element-wise approach with the help of the isogeometric subdivision, which was defined for this purpose (cf. Def. 4.1 and Lem. 4.2). Furthermore in Sec. 4.4.1 we have studied which basis functions are non-zero over an element and that they also obey a tensor product structure. On each element we have to take into account these local basis functions and compute their global numbers for the global stiffness matrix.

Just like in FEM we can use Gaussian quadrature shown in Fig. 4.12 on each element to compute the integrals defined over it. It suggests itself to extend the one-dimensional quadrature rules via tensor product just like in Fig. 4.12. An alternative approach to quadrature is the use of macro-elements, shown in [59]. The main idea here is to make use of the high inter-element regularity of the splines and employ fewer quadrature rules that are based on several adjacent elements. This approach emphasizes the regularity and the size of the basis functions and not the interplay between the supports as mentioned in Sec. 4.2.2. Macro-elements consist of several isogeometric elements, but due to the possible occurrence of knot domains with empty interior several configurations

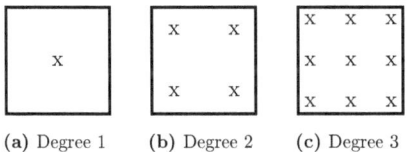

(a) Degree 1   (b) Degree 2   (c) Degree 3

Figure 4.12: Gaussian quadrature points

of the macro-element may be necessary. Furthermore, combining macro-elements with local refinement (see Chap. 5) creates difficulties and we will avoid this here.

Since we know the indices of the element as well as the number of the basis functions, we can evaluate the latter at the quadrature points. The evaluation of B-spline was discussed in Sec. 2.4.1 and Alg. 2.1 and it was shown that all B-splines that are nonzero at the point are evaluated at once. NURBS values are computed with consideration of the weights out of the B-spline evaluations.

The general assembly procedure is very similar to FEM (see Alg. 3.2). For ease of notation we define the integrand $\alpha(\varphi_i, \varphi_j, G)$ such that

$$a(\varphi_i, \varphi_j)|_{T_k} = \int_{T_k} \alpha(\varphi_i, \varphi_j, G)\, du. \tag{4.29}$$

For each element we evaluate the bilinear form with two basis functions with the help of the quadrature rule and the transformation given by the geometry mapping G. Furthermore the integral domain is the element T_k itself, which is always a rectangle. Therefore any further transformations (e.g to some kind of reference element) are not necessary. For example for the Poisson equation we have

$$\alpha(\varphi_i, \varphi_j, G) = (DG^{-T} \nabla \hat{\varphi}_i) \cdot (DG^{-T} \nabla \hat{\varphi}_j) |\det DG|. \tag{4.30}$$

Finally, all this is summarized in Alg. 4.1.

Algorithm 4.1 Basic isogeometric stiffness matrix assembly algorithm

---
for element index k **do**
  for local basis index i **do**
    for local basis function j **do**
      $a_{\text{loc}} = 0$
      for all quadrature points q in T_k **do**
        $a_{\text{loc}}{+}= w_q\alpha(\varphi_i, \varphi_j, G))|_q$
      end for
      $a_{g(j),g(i)} = a_{\text{loc}}$
    end for
  end for
end for

Pre-evaluation

Isogeometric analysis does not allow a similar concept like the reference element concept to e.g. create element stiffness matrices that only need to be computed once, because of the complexity of the geometry mapping and the fact that the basis functions are affected by the knot domains in their support extension and not only by one element. Also the isogeometric reference element discussed in Sec. 4.4.1 does not resolve these circumstances. Nevertheless, we still have the opportunity to make use of similar techniques. For a multivariate B-spline function

$$B_{ij}(u, v) = B_i(u)B_j(v) \tag{4.31}$$

we compute the function value by evaluating univariate B-splines. Also, derivatives are products of univariate function values as seen in Sec. 2.3. Since the evaluation points are known by the mesh structure and the quadrature rule we can make use of the tensor product structure and compute all univariate values beforehand. So if we assume that we have to evaluate the B-splines on a $n \times m$ grid, instead of doing $n \cdot m$ single evaluations we only need to carry out $n + m$ evaluations beforehand. These values can also be used for Neumann boundary conditions, although the effect is less remarkable, there are much fewer evaluation points compared to the assembly of the stiffness matrix.

For NURBS the product of the basis functions in each parameter coordinate is replaced by

$$N_{ij}(u, v) = \frac{B_i(u)B_j(v)\omega_{ij}}{\sum_{i,j}^{n_i,n_j} B_i(u)B_j(v)\omega_{ij}}. \tag{4.32}$$

The difficulty here is that the NURBS are still influenced by the geometry mapping through the weights ω_{ij}. So it would not be reasonable to compute the univariate values over the whole mesh. Instead, we take the precomputed B-spline values and compute the NURBS values within the assembly loop at once for all evaluation points in an element. Similarly, we can preevaluate all values of the Jacobian and the transformation matrices of the geometry mapping for all quadrature points in an element in the same manner. Then these costly values do not need to be computed multiple times. The modified assembly algorithm is sketched in Alg. 4.2.

4.6.4 Boundary Conditions

Implementing Neumann boundary conditions into isogeometric analysis is not different from FEM (see Sec. 3.5.3). Instead of integrating over the boundary faces, we compute the integral over boundary curves or surfaces. At these points we always are at the end of the interval of one parametric coordinate and therefore the B-splines and NURBS are equal to one. For example, if the boundary curve Γ_N is parameterized by $\boldsymbol{G}|_{u \in [a,b], v=0}$, we can formulate the integral as

$$\int_{\Gamma_N} g(x, y) \, B_{ij} \circ \boldsymbol{G}^{-1}(x, y) \, dx = \int_{u \in [a,b]} g(\boldsymbol{G}(u, 0)) \, B_i(u)B_j(0) \, dx =$$
$$= \int_{u \in [a,b]} g(\boldsymbol{G}(u, v)) \, B_i(u) \, dx. \tag{4.33}$$

Algorithm 4.2 Assembly algorithm with preevaluation

compute univariate basis values φ_i at quadrature points
for element index k **do**
  compute NURBS basis values (if necessary)
  compute geometry transformations on T_k
  for local basis index i **do**
    for local basis function j **do**
      $a_{\text{loc}} = 0$
      for all quadrature point q in T_k **do**
        look up basis values
        look geometry transformations
        add up to a_{loc}
      end for
      $a_{g(j),g(i)} = a_{\text{loc}}$
    end for
  end for
end for

So the integral only depend on an univariate B-spline. This can be applied to NURBS and higher dimensions analogously.

Also Dirichlet boundary conditions work similar to FEM. If we want to prescribe values to certain degrees of freedom we can use the same techniques discussed in Sec. 3.5.3. The challenge here is to specify the values. It is not possible to simply use function evaluations here, because spline basis function are not interpolatory at the nodal variables except in the corners of the parametric domain. So instead we have to employ a curve or surface fitting to find the coefficients for a good approximation. Only in the simplest cases like piecewise constant functions we can simply set all the coefficients to this value, because of the partition of unity property.

The task of finding the appropriate coefficients for the Dirichlet boundary conditions is not one-of-a-kind for isogeometric analysis. Especially meshless or generalized FEM methods share this property. We refer to [9] for a discussion and further solution approaches.

4.6.5 Visualization

Visualization of geometric as well as numerical data is of great importance. In the simplest case in isogeometric analysis the simulation result itself again is a shape. This may for example occur in simulation of problems in elasticity, where we want to visualize a deformed shape. Let the inital shape be given by

$$G(u, v) = \sum_{i,j} N_i(u) N_j(v) P_{ij}, \qquad (4.34)$$

and $\boldsymbol{u}_{ij}$ the part of the solution vector that corresponds to the indices i and j. Then we can represent the deformed shape by

$$\boldsymbol{G}_{\text{def}}(u, v) = \sum_{i,j} N_i(u) N_j(v) (\boldsymbol{P}_{ij} + \boldsymbol{u}_{ij}) \tag{4.35}$$

and visualize it like the geometry with new control points. This is done, for example, for fluid-structure interaction problems (see [54]).

Other solution fields like the stress tensor or scalar measures of stress are not representable like this. Instead we directly evaluate the desired quantities with the algorithms that are used for evaluating the geometry.

From the finite element point of view this is also unfamiliar, because we have no underlying mesh where the data belongs to. This is replaced by the geometry representation and evaluation of the linear combination of the basis functions. So typically an evaluation grid is chosen and the values of interest are computed on each of these points. An important algorithmic aspect here is that we can make use of grid evaluation algorithms that compute a spline at a great number of points efficiently. Some visualizations are shown in Sec. 4.7 and 5.5.

4.7 Simulation Examples

In this section we will present some numerical examples, which show the capabilites of isgeometric analysis. Further examples are found in Sec. 5.5 with a focus on local refinement introduced in Chap. 5.

4.7.1 Heat Conduction on a Half-Disk

We start with the Poisson problem introduced in Sec. 3.1.1.

$$-\Delta u = f \tag{4.36}$$

on the upper half-disk with the right hand side

$$f(r, \varphi) = 6r^{\frac{1}{2}} \sin(\frac{\varphi}{2}) \tag{4.37}$$

given in polar coordinates (r, φ). The geometry and the boundary conditions are shown in Fig. 4.13a. We employ zero Dirichlet as well as zero Neumann boundary conditions. By simple calculation we can show that the exact solution of this problem is

$$u(r, \varphi) = (1 - r^2) r^{\frac{1}{2}} \sin(\frac{\varphi}{2}), \tag{4.38}$$

again given in polar coordinates.

To represent the half-disk correctly we need a rational spline of least degree two in one parameter direction. Still, the other one direction may only have degree one, but we will use p-refinement in order to have same degrees. So we obtain a parameterization with degree two in each direction and a collapsing side at the orgin. The 15 initial

(a) Boundary conditions: Dirichlet Γ_D and Neumann Γ_N (b) Control grid

Figure 4.13: Heat equation on half-disk

control points are shown in Fig. 4.13b. For the simulation we furthermore apply two h-refinement steps to have a suitable starting mesh. With the use of NURBS the geometry and especially the half circle is represented exactly even at lowest level.

Due to the adjacency of the Dirichlet and the Neumann boundary this solution is not within $H^2(\Omega)$ and therefore the global error estimate in Th. 4.8 is not applicable. Instead, if we solve the problem numerically, we observe a reduced convergence rate as shown in Fig. 4.14.

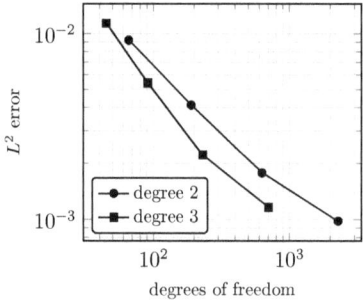

Figure 4.14: Convergence plots for the Poisson problem on the half-disk

4.7.2 Heat Conduction on an L-shape

We now want to look at a simulation example with two different parameterizations. The problem consists of the Laplace equation on an L-shape domain $\Omega = [-1, 1]^2 \setminus [0, 1]^2$ with boundary conditions shown in Fig. 4.15. These are adapted to the analytical solution given by

$$u(r, \varphi) = r^{\frac{2}{3}} \sin\left(\frac{2\varphi - \pi}{3}\right) \tag{4.39}$$

in polar coordinates. Again the solution is not H^2 continuous over the computational domain due to the region near the reentrand corner and therefore the convergence estimates in Sec. 4.3 do not hold.

For the L-shape we have two different parameterizations available. The first has a C^0 edge from the reentrand corner to its opposite corner. The 15 control points

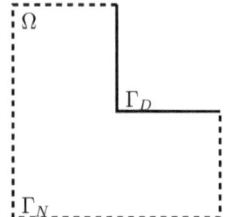

Figure 4.15: Boundary conditions for the L-shape problem: Dirichlet Γ_D and Neumann Γ_N

are shown in Fig. 4.16a. The second one is a C^1 parameterization and its 12 control points are shown in Fig. 4.16b. We see that control points coincide in the lower left corner and therefore the paramterization is not regular any more. Nevertheless we will see that we still obtain reasonable results and therefore this is one example for the paramteric violations discussed in Sec. 4.2.3. For both parameterizations we apply two h-refinement steps to get a initial mesh for the simulation.

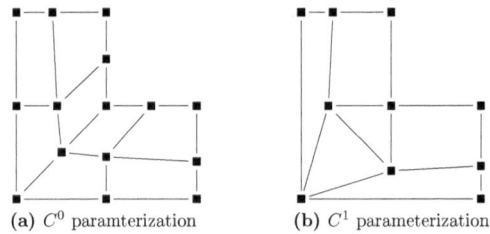

(a) C^0 paramterization   (b) C^1 parameterization

Figure 4.16: Control point grid for L-shape parameterizations

In Fig. 4.17 the convergence plots for the L-shape for both paramterizations are shown. Whereas both parameterization have the same convergence rate, which may be lead back to the lacking continuity of the solution, we see that the C^1 parameterization achieves a better overall approximation.

4.7.3 Plate with a Hole

The second example is an infinite plate with a circular hole under in-plane tension in x-direction (also see [58, 35]). We study the stationary linear elasticity problem

$$\operatorname{div} \boldsymbol{\sigma}(\boldsymbol{u}) = \boldsymbol{f} \tag{4.40}$$

under the assumption of plane stress (see Sec. 3.1.2). Due to symmetry the computational domain is restricted to a quarter. Furthermore we employ Dirichlet boundary conditions according to the analytical solution of the infinite plate at the boundary of our finite computational domain, see Fig. 4.18a.

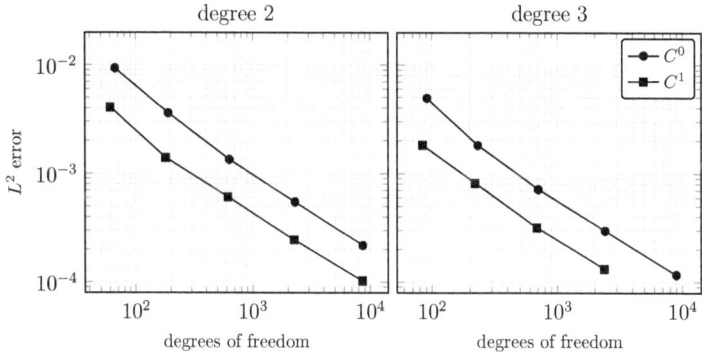

Figure 4.17: Convergence plots for the Poisson problem on the L-shape for different parameterizations and different ansatz degrees

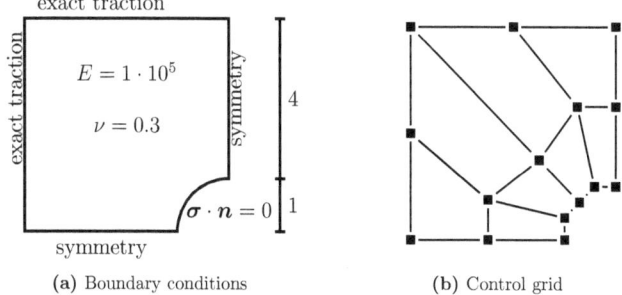

(a) Boundary conditions   **(b)** Control grid

Figure 4.18: Plate with hole

The exact stress distribution for this problem (see e.g. [51]) is given by

$$\sigma_r = \frac{T_x}{2}(1 - \frac{R^2}{r^2}) + \frac{T_x \cos(2\theta)}{2}(\frac{3R^4}{r^4} - \frac{4R^2}{r^2} + 1,) \tag{4.41a}$$

$$\sigma_\theta = \frac{T_x}{2}(1 + \frac{R^2}{r^2}) - \frac{T_x \cos(2\theta)}{2}(\frac{3R^4}{r^4} + 1), \tag{4.41b}$$

$$\sigma_{r\theta} = \frac{T_x \sin(2\theta)}{2}(\frac{3R^4}{r^4} - \frac{2R^2}{r^2} - 1). \tag{4.41c}$$

Again we represent the geometry exactly, especially the circular hole, by making use of NURBS. Therefore we employ a bivariate spline parameterization of degree two in each direction with 15 control points, which are shown in Fig. 4.18b.

For the simulation we want to look at different ansatz degrees up to four and therefore apply p-refinement first. For the lowest degree two we also apply a h-refinement step to get a suitable starting resolution. Figure 4.19a shows the convergence behavior of the first principal stress component in the L^2 norm. If we compare them, we see that the convergence rate increases with higher degree.

The condition number in the 2-norm for the stiffness matrices is also plotted in Fig 4.19b. As expected the condition number increases with raising degrees of freedom, but it can be seen that there is not significant difference between the degrees.

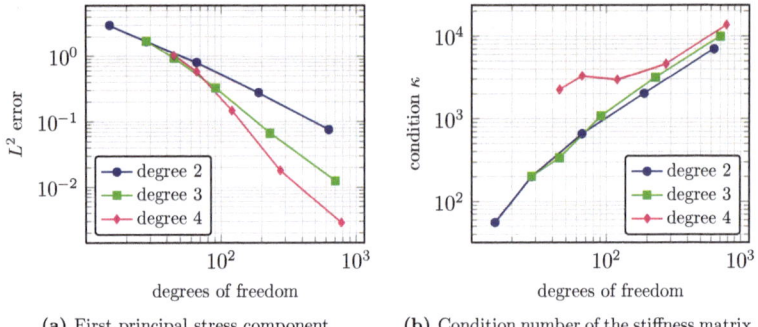

(a) First principal stress component (b) Condition number of the stiffness matrix

Figure 4.19: Results for plate with hole problem

4.7.4 Turbine Blade

Until now we have only treated static problems. For the extension to dynamic cases and conduct a modal analysis we look at Eq. (3.17) and include the time derivative, which resembles Newton's second law. This leads to

$$\rho\frac{\partial^2}{\partial t^2}\boldsymbol{u} = \text{ div } \boldsymbol{\sigma} + \boldsymbol{f}. \tag{4.42}$$

We are especially interested in the natural frequencies and modes and assume $\boldsymbol{f} = 0$. Discretising this differential equation analogously as in the sections before, we obtain

$$\boldsymbol{M}\frac{\partial^2}{\partial t^2}\boldsymbol{u}(\boldsymbol{x},t) + \boldsymbol{A}\boldsymbol{u}(\boldsymbol{x},t) = 0 \tag{4.43}$$

with the mass matrix

$$(\boldsymbol{M})_{ij} = (\rho\varphi_i, \varphi_j). \tag{4.44}$$

Employing the ansatz $\boldsymbol{u}(\boldsymbol{x},t) = \boldsymbol{u}(x)e^{i\omega t}$ this results in

$$(-\omega^2\boldsymbol{M} - \boldsymbol{A})\boldsymbol{u}(x)e^{i\omega t} = 0. \tag{4.45}$$

This is a generalized eigenvalue problem

$$\boldsymbol{A}\boldsymbol{u} - \lambda\boldsymbol{M}\boldsymbol{u} = 0 \tag{4.46}$$

which can be solved to obtain the eigenmodes and frequencies.

We will apply this technique to a model inspired by real-life applications. The geometric model of a turbine blade was created at the Institute for Applied Geometry at the University Linz and is described in the report [87]. More details about the creation process of such a geometry are found in [3]. The parameterized blade created this way is shown in Fig. 4.20 and consists out of 81 control points.

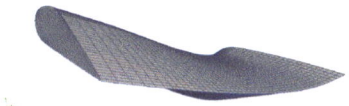

Figure 4.20: Geometric model of the turbine blade

We fix the blade at the right side, apply h-refinement and compute some eigenforms with the previous mentioned approach. Some results are shown in Fig. 4.21 for a model with 192 degrees of freedom. For further information about modal analysis in isogeometric analysis we refer to [36].

Figure 4.21: Eigen forms of the turbine blade

4.7.5 Wheel Disk

An interesting application for isogeometric analysis is the wheel disk. The side of the wheel that touches the rail is defined by techncial specifications and is defined through a given spline curve. The cross section through a wheel is shown in Fig. 4.22a. The parameterization of the wheel disk shown in Fig. 4.22b was created in the diploma thesis [100] with the techniques shown in [3].

The simulation results achieved with isogeoemetric analysis are shown in Fig. 4.23. A force is applied at the bottom of the wheel, whereas the axis hole is fixed. We observe the increase of stress in the region between the applied force and the mount.

We want to do a simple runtime analysis for this three-dimensional problem and look at three different cases: starting from an initial configuration with degree two and 3690 degrees of freedom we apply uniform h-refinement to get the second case with 21546 degrees of freedom. The third is generated out of the first by p-refinement, so that we have degree three and 13230 degrees of freedom. The run time for each case is shown in Tab. 4.2. All computations were done on a standard PC (Intel Core i7 920 with 3 GB DDR3 RAM) with a single process only. We see that the assembly

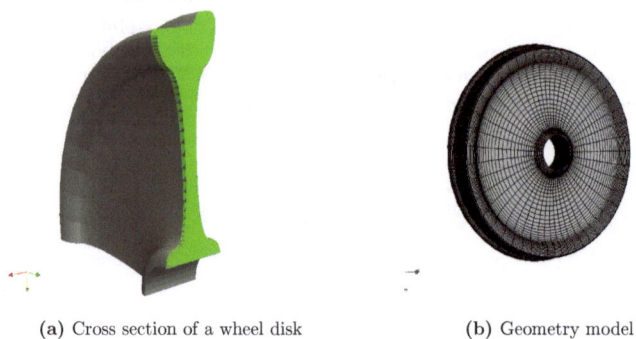

(a) Cross section of a wheel disk   (b) Geometry model

Figure 4.22: Wheeldisk

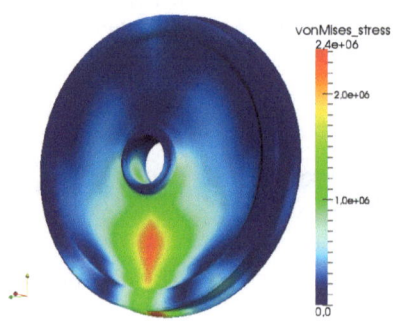

Figure 4.23: Simulation result for the wheel disk: von Mises stress

takes most of the computation time, but its percentage decreases under h-refinement, because also solving the linear equation gets more involving. The third case shows although the number of degrees of freedom is clearly less than in the second case we need similar run time for the assembly due to the higher degree of the basis functions. The postprocessing is the same for all three cases and takes the same time. Nevertheless as it was discussed in Sec. 4.6.5 the evaluation grid can be choosen independently from the isogeometric mesh. Therefore the runtime for postprocessing mainly depends on the desired resolution for the visualization.

Table 4.2: Comparison of computation time for wheel disk example

		initial	h-ref	p-ref
	degree	2	2	3
	dof	3690	21546	13230
run time [s]	assembly	404	3254	3943
	solving	16	709	249
	postprocessing	7	7	9
	total	427	3971	4201
percentage	assembly	94.6%	81.9%	93.8%
	solving	3.7%	17.7%	5.9%
	postprocessing	1.6%	0.17%	0.21%

4.8 Conclusion and Applications

Before continuing with the next chapter and discussing a special aspect of isogeometric analysis it is a good time to shortly summarize the results so far and to especially mention applications of this method.

Geometry Representation and Generation   We have seen that isogeometric analysis creates a link between the simulation framework and the CAGD geometry. There is no standard grid generation (see e.g. [24]), which can turn out to be very involving and requires a lot of computational effort (see e.g. [38]). Through the use of a parameterization as an integral part we can ensure the exact representation at all refinement levels. Of course, isogeometric parameterization and geometry generation is still in an infant stage, but nevertheless first results and techniques haben been obtained (see [3]) and were used to create the three dimensional models in Sec. 4.7.

Information Exchange   Due to the unified description of geometry and computational mesh in isogeometric analysis the information exchange is simplified significantly. Especially for shape optimization, which aims at improving a given geometry with respect to certain quantities that are typically results from a numerical simulation, this connection is very favorable. Since the optimization is computationally costly, the discretization for the pure simulation is replaced by a smaller model, which results in further efforts and transfer of information. In isogeometric shape optimization (see e.g. [98, 70]) the optimization as well as the numerical simulation can be executed on the common NURBS description. Also for multifield problems like fluid-structure interaction promising results have been obtained (see [54, 20]).

Simulation Features   From the point of view of numerical analysis and simulation we have basis functions of theoretically arbitrary smoothness without the difficulties of managing derivatives and very high degrees as for Hermite elements. This is very favorable for higher order problems that usually need high smoothness. With isogeometric analysis it is e.g. possible to employ shell models that require e.g. $\mathcal{C}^1$ continuity (see [63]). Furthermore, these objects are often given as free-form surfaces, which also fits well into isgeometric analysis.

The higher smoothness is also favorable if differentiated values like stress, which may be more interesting from an engineering points of view than the displacements, are computed in a postprocessing step. In order to illustrate this we show an example from [97], where a solid cylinder under load was investigated. The stress distribution, computed with isoparametric FEM and isogeometric analysis, along the boundary of cylinder base is shown in Fig. 4.24. We see that the results for isogeometric analysis are smoother and do not suffer from jumps.

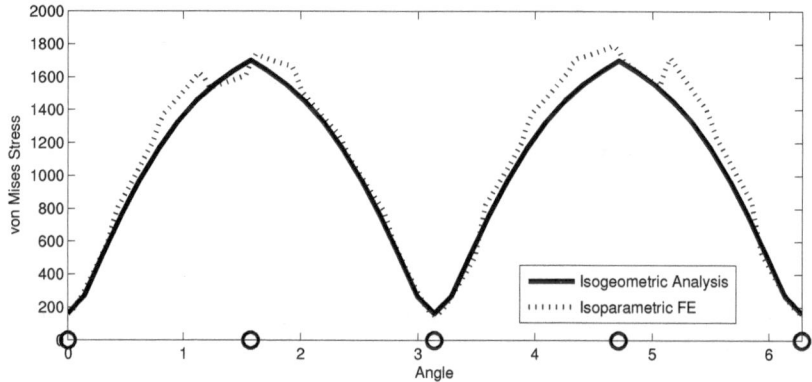

Figure 4.24: Stress distribution at the bottom of a solid cylinder

The influence on the numerical algebra is also evident. Due to the increased overlap the resulting matrices are not only less sparse, but also smaller in size, because typically lesser degrees of freedom are needed.

Moreover, the refinement is also given through the CAGD routines without change of the boundary. In finite elements the refinement implementation has to take special care in case of curved boundary (e.g. see [79]).

Conclusion Isogeometric analysis tackles the gap between FEM and CAGD with respect to representation and information interchange. Furthermore the higher smoothness of the basis function has some positive effects on the simulation. Nevertheless, there is one aspect that does not benefit from this. Local refinement is not possible in isogeometric analysis presented so far. The following chapter will discuss this further and offers an approach to resolve this problem.

Chapter 5

Local Refinement for Isogeometric Analysis

Things alter for the worse spontaneously,
if they be not altered for the better designedly.

(Francis Beacon)

During the last chapter several aspects of isogeometric analysis were discussed. This chapter will focus on the main topic of this thesis: adaptivity. Adaptive simulation is essential for simulating complex problems and using the available resources where they are needed. The uniform refinement techniques presented in the previous chapter turns out not to be flexible enough. In this chapter we will strongly rely on the previous section about isogeometric analysis and expand the concepts introduced therein to introduce an effective method for local refinement. The approach we propose is based on function spaces from CAGD, which were not especially designed for this purpose. We transform this idea to isogeometric analysis by adjusting it to create a basis with desirable properties, but also add an element concept that is suitable for the simulation setting and extendable to more advanced techniques like error estimation. It is very crucial at this point that we have taken both aspects, applied geometry and numerical analysis, into account.

We will proceed as follows: point of departure will be an overview and discussion of different refinement techniques and their properties. Then the main focus will be hierarchical B-Splines and their properties. The next section will transfer this idea into the context of isogeometric analysis and combine it with the concepts introduced in the previous chapter. We will define several hierarchical structures and investigate their relationships. Promising numerical results will be shown at the end.

Again we will discuss everything in two dimension out of readability. Nevertheless everything works analogously for more dimensions.

The main ideas of hierarchical refinement in this chapter can alternatively also be found in [95] with some differences in notation and focus.

5.1 Introduction to Adaptive Simulation

For an adaptive simulation the simple uniform refinement procedure is replaced by two more sophisticated components:

- a local error estimate to evaluate the quality of the solution and determine the region with higher errors,

- an automatic local mesh refinement technique to adapt the mesh accordingly.

Typically the first point is fulfilled by a local a posteriori error estimate which was already introduced in Sec. 3.4.4. The second one was mentioned for FEM in Sec. 3.4.3: the finite elements are subdivided and the whole mesh is further altered to obtain an admissible subdivision or further measures to take care of hanging nodes are invoked.

In isogeometric analysis the normal refinement procedure can not be applied to generate a local increase of resolution. An example with an L-shape geometry is shown in Fig. 5.1. If we want to refine the neighborhood of the reentrant corner by h-refinement, the refinement propagates along the knot lines. So unlike in FEM we do not even have

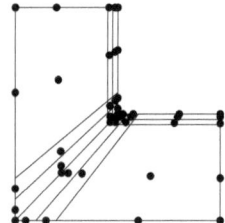

Figure 5.1: Tensor product prevents local refinement

a chance to apply standard knot insertion to a single element only and to achieve local refinement. Also B-Splines and NURBS carry some difficulties. The bigger support of each basis function and the resulting overlapping creates a more complicated structure. Also the property of higher continuity is difficult to control and maintain with local approaches.

As it was discussed in Sec. 4.2 isogeometric analysis consists of three components: a geometry mapping, an element structure and the basis functions. Even if we assume that we are able to leave the geometry mapping and the elements untouched and just alter the basis functions, we get an approach that is related to so called "basis function refinement" (cf. [52]), where the main focus is on subdividing the functions. Unfortunately this neglects the mesh structure underneath and, for example, does not answer the question of efficient assembly and local error estimation. Therefore it is likewise important to adopt the mesh structure accordingly.

5.1.1 Aspects of Refinement in Isogeometric Analysis

Before discussing individual refinement schemes we want to identify which properties may be helpful for local refinement. We start with properties that are known from finite element methods.

locality The refinement should be restricted to a certain area. In finite elements this is an opposing requirement to admissible subdivisions and has to be ensured by a suitable strategy on how to refine each element.

linear independence The created basis functions should not lose their inherent property of being linearly independent, which would result otherwise in bad numerical behavior.

nested spaces This is the defining property of refinement, because we still want the old representation within the new enlarged space. Due to this property it is assured that the error does not increase.

partition of unity The partition of unity property is very typical for FEM and generalizations like partition of unity FEM or meshless methods (cf. [73], [9]).

special case uniform refinement If the refinement scheme is applied globally, this results in the uniform refinement technique.

There are some properties that are unique to isogeometric analysis and nevertheless should be considered as well.

representability of geometry This is one major feature of isogeometric analysis. It assures that linear solutions are representable and the geometric description does not change under refinement (see Sec. 4.2.4).

unified representation This property means that all the information is found in one underlying structure. For example, the basis function are computed over one mesh structure. This has major influence of the data structures for the implementation.

suitable for any spline space The refinement technique can be used for any arbitrary degree. Moreover, multiple knots and therefore changing continuity are not problematic.

We will discuss several refinement techniques with respect to these properties in the next section.

5.1.2 Isogeometric Analysis Refinement Techniques

Several proposals were made how to achieve local refinement in isogeometric analysis. We will shortly review some of them and give references for more details.

Multipatch

One basic idea to create local refinement is to subdivide the parameter domain into several patches and refine a patch uniformly for itself. We therefore employ a multipatch model introduced in Sec. 4.2.2. This approach does not change how the basis functions are obtained, but the mesh they are defined on and can be seen as h-refinement on a macro-scale. Instead of describing the geometry with different patches, we subdivide the given parameterization and use standard refinement on the patches. Hereby connecting functions on the patches is the major difficulty as mentioned before. By identification of control points we just obtain C^0-continuity between the blocks. For higher smoothness we have to apply further constraints on the basis functions. Similar to hanging nodes in finite elements also the different mesh sizes are treated with constraints. Furthermore, it is difficult to dynamically adapt the refinement region, for example, change its size or refine over several levels, because the subdivision of blocks cannot be easily changed or moved without a lot of computation. This idea was already discussed in [35] for isogeometric analysis and [60, 61] for B-Spline FEM.

T-splines

T-splines introduced in [85] and [84] are a CAGD technique, that tackles the problem of too many unneeded control points in a geometric model. The use of T-splines for refinement in isogeometric analysis was introduced in [42] and [19].

The basic idea of T-splines is based on the fact that a NURBS function N_i^p only depends on the knots $u_i, \ldots, u_{i+p+1}$ (see B-spline Def. 2.7). In two dimensions and for degree three we get a symmetric stencil with center knots shown in Fig 5.2a. The stencil is applied on a so called T-mesh, a structure with its own rules and that breaks the tensor product structure, to obtain the knots and then use the standard definition to define the NURBS.

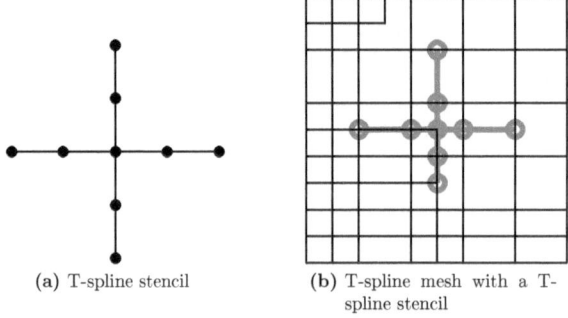

(a) T-spline stencil   (b) T-spline mesh with a T-spline stencil

Figure 5.2: T-splines

There are two different viewpoints on T-splines: in CAGD T-splines offer saving control points by removing them from a NURBS control mesh. From the simulation point of view we do not start from a fine grid, but refine a coarse grid instead. In the latter it may occur that T-spline refinement propagates along the knot lines (see [42]). The reason for that is the dependency on the recursive evaluation formula. Furthermore, it was shown in [28] that there is a configuration where the T-spline functions become linear dependent. Also T-splines are initially only defined for degree three. For a generalization the algorithms would have to be adopted heavily. For example the use of an even degree would also disturb the symmetry of the stencil, which would then lead to an undesired anisotropic behavior. Nevertheless, efforts were made to further adapt the T-splines to be more suitable for isogeometric analysis and apply further restriction to enforce the linear independence, which is described in [68] and [83]. Also the combination with Bézier extraction mentioned in Sec. 4.4.2 is discussed in [82].

HTP-splines

Hierarchical tensor product splines (HTP-splines) were also introduced as a CAGD technique in [40]. Although the name suggest some relationship with hierarchical B-splines this method is very different and for example does not employ B-splines or NURBS as basis functions.

Starting point is a locally refinable mesh that only consists of quadrilaterals but allows hanging nodes. Over each quadrilateral a polynomial is defined with the help of function values and derivatives. This is strongly related to Hermite polynomials, but also involves the restriction concerning degree and continuity. So the functions are bound to C^1 continuity and have degree three at most. This, of course, influences the geometry mapping G. If we want G to be representable in the ansatz space it has to be restricted to C^1 functions as well.

From the finite element point of view this method is a variant of a mapped Hermite element approach with hanging nodes. The details about the numerical realization and application in isogeometric analysis are found in [74].

Summary

A short summary of the refinement techniques discussed in the previous section is shown in Table 5.1. We see that it is important to look at all components of isogeometric analysis. Each method we have presented has its special emphasize. Multipatch refinement only focuses on a new mesh, but ensuring the smoothness of the basis functions turns out to be very involving. T-splines are special basis functions, but their properties and mesh refinement behavior need further investigations and adjustments. HTP-splines replace the NURBS basis functions at all, but this also effects the geometry representation.

Table 5.1: Summary: refinement techniques

Component	Multipatch	T-spline	HTP-spline
Basis function	NURBS	T-splines	piecewise C^1 polynomials
Mesh	Multipatch	T-mesh	quadrilateral T-mesh
Geometry mapping	NURBS	NURBS	piecewise C^1 polynomials

The properties discussed in Sec. 5.1.1 for the refinement techniques are summarized in Table 5.2. Uniform B-spline refinement behaves very well, except its major drawback that it is not local. Multipatch refinement has its main difficulties in maintaining the continuity across the patches. Local refinement and linear independence still need to be enforced by T-splines and this method is only applicable for degree three. HTP-splines are also limited to C^1 continuous functions.

Table 5.2: Refinement techniques and their properties

	local	linear independence	nested	partition of unity	special case uniform	geometry representation	unified representation	arbitrary degree/continuity
uniform B-spline Refinement	✓	✓	✓	✓	✓	✓	✓	✓
Multipatch	✓	✓	✓	✓	✓	✓		
T-splines			✓	✓	✓	✓	✓	
HTP-splines	✓	✓	✓	✓				
hierarchical B-Splines	✓	✓	✓	O		✓	✓	✓

We now want to investigate an approach, hierarchical refinement, that leaves the geometry function unaltered, but ensures suitable behavior for the basis functions as well as the mesh. In particular, hierarchical refinement is local, ensures linear independence and is applicable for any degree or continuity. Other properties for the hierarchical refinement are also stated here for completeness. They will be all discussed in detail in Sec. 5.3.2.

5.2 Hierarchical B-Splines

In this section we want to look at basis functions that we will use later on for our refinement technique — hierarchical B-splines and their modifications. We start this investigation by stating the definition of the spaces and discuss their main properties.

5.2.1 Basic Definitions

We define a hierarchical structure used several times in the following. Let $(\mathcal{S}^\ell)_{\ell=1,\dots,L}$ be a finite nested sequence of L bivariate B–spline spaces

$$\mathcal{S}^1 \subset \mathcal{S}^2 \cdots \subset \mathcal{S}^L, \tag{5.1}$$

defined over the parametric space Ω_0 and spanned by B-splines $\mathcal{B}^\ell$. We assume that the basis functions get "smaller" with each $\mathcal{S}^\ell$. This suffices for the time being and we will be more concrete later on. Furthermore we define a finite sequence of L bounded closed sets $(\Omega^\ell)_{\ell=1,\dots,L}$

$$\Omega^L \subseteq \Omega^{L-1} \subseteq \cdots \subseteq \Omega^1 = \overline{\Omega}_0 \tag{5.2}$$

which resembles the refinement regions.

Classical Hierarchical B-Splines

Hierarchical B-splines were first introduced by Forsey and Bartels in [47]. In this approach overlays of hierarchical controlled subdivisions are used. The setting is as follows: at each level ℓ we select the functions from $\mathcal{B}^\ell$ with support in Ω^ℓ

$$\mathcal{B}^\ell_{loc} := \{\varphi \in \mathcal{B}^\ell \,|\, \operatorname{supp} \varphi \subseteq \Omega^\ell\} \tag{5.3}$$

and define a geometric object S^ℓ with respect to this selection

$$S^\ell = \sum_{\varphi \in \mathcal{B}^\ell_{loc}} \varphi \boldsymbol{P}_\varphi \tag{5.4}$$

and with given control points $\boldsymbol{P}_\varphi$. S^1 is the initial object and the $S^2, \ldots S^L$ are local overlays with support within Ω^ℓ, respectively, that describe additional features. The overall object is given by

$$S = \sum_{\ell=1}^{L} S^\ell. \tag{5.5}$$

Created for application in CAGD the representation of S does not need to be unique. The reason for this is that the hierarchical B-splines do not need to be linearly independent. The functions for describing the overlays were just added and already existing functions may get several representation possibilities.

Linearly Independent Multilevel B-Splines

Based on the idea of hierarchical splines we now want to construct a basis and therefore preserve linear independence as well as the local refinement property. Instead of only adding functions we also have to remove functions. This approach was investigated in [64] and [65] and we will make some slight extensions to this.

We first concretise how the nested spaces are related to each other. Let $(\boldsymbol{U}^\ell, \boldsymbol{V}^\ell)$ be the knot vectors of the basis $\mathcal{B}^\ell$ that spans the space $\mathcal{S}^\ell$. We will focus on h-refined spaces here and therefore the degrees (p_u, p_v) are supposed to be equal for all levels, but p_u is allowed to differ from p_v. We assume that the knot vectors are nested, that means that for any knot and its multiplicity m^ℓ at level ℓ holds

$$m^\ell \leq m^{\ell+1}, \; \ell = 0, \ldots, L-2 \tag{5.6}$$

where we set the multiplicity of a knot that is not present in the knot vector to be zero. At each level the boundary $\partial\Omega^\ell$, $\ell = 1, \ldots, L$, is aligned with the knot lines of $\mathcal{S}^{\ell-1}$.

By slightly generalizing the selection mechanism for the underlying tensor product B–spline bases $\mathcal{B}^1, \ldots, \mathcal{B}^L$ introduced in [64], the basis $\mathcal{A}$ for the hierarchical spline space is defined as follows.

Definition 5.1. *Let $(\Omega^\ell)_{\ell=1,\ldots,L}$ be a sequence of bounded closed sets and $(\mathcal{B}^\ell)_{\ell=1,\ldots,L}$ a sequence of B-spline bases as stated above. Then the hierarchical basis $\mathcal{A}$ is recursively constructed as described below.*

- *Initialization:*

$$A^1 = \mathcal{B}^1. \tag{5.7}$$

- *Construction of $A^{\ell+1}$ from A^ℓ*

$$A^{\ell+1} = (A^\ell \backslash A_\ominus^\ell) \cup A_\oplus^{\ell+1}, \qquad \ell = 1, \ldots, L-1, \tag{5.8a}$$

where

$$A_\ominus^\ell = \{\varphi \in A^\ell : \operatorname{supp}\varphi \subseteq \Omega^{\ell+1}\}, \tag{5.8b}$$

and

$$A_\oplus^{\ell+1} = \{\varphi \in \mathcal{B}^{\ell+1} : \operatorname{supp}\varphi \subseteq \Omega^{\ell+1}\}. \tag{5.8c}$$

- $\mathcal{A} = A^L$.

Starting from the B-spline basis $\mathcal{B}^0$ we iteratively take into account the next level. With the set $A_\ominus^\ell$ we remove all basis functions of the previous level whose support is contained in $\Omega^{\ell+1}$ and then with $A_\oplus^{\ell+1}$ replace them by refined basis functions in $\mathcal{B}^{\ell+1}$ that cover $\Omega^{\ell+1}$. This interpretation is substantiated by

Corollary 5.2. *Let the prerequisites from Def. 5.1 hold. Then*

$$\operatorname{span} A_\ominus^\ell \subseteq \operatorname{span} A_\oplus^{\ell+1}, \qquad \ell = 1, \ldots, L-1. \tag{5.9}$$

Proof. Let $\varphi \in \operatorname{span} A_\ominus^\ell$. Then especially $\varphi \in A^\ell$ and therefore φ is representable in the basis $B^{\ell+1}$ due to the nested sequence of spline spaces. All coefficients of the basis functions without support within $\Omega^{\ell+1}$ have to be zero due to the local linear independence and therefore $\varphi \in \operatorname{span} A_\oplus^{\ell+1}$. $\qquad\square$

Definition 5.1 is adopted from [95] with slight reformulations. Here we have just included the setting we will use further in Sec. 5.3. A more general setting is elaborated in the article.

We can directly conclude a characterization if a basis function belongs to the hierarchical basis from its definition.

Corollary 5.3. *For $\varphi \in \mathcal{B}^\ell$ it holds*

$$\varphi \in \mathcal{A} \Leftrightarrow (\varphi \in A_\oplus^\ell \wedge \varphi \notin A_\ominus^\ell) \tag{5.10}$$

with special case $A_\oplus^0 := \mathcal{B}^0$.

A one-dimensional example for the linear independent hierarchical basis is shown in Fig. 5.3. At the top, quadratic B-splines are plotted for a given equidistant knot vector. The plot in the middle displays the B-splines after each knot span has been subdivided by knot insertion, the next level. The aim now is to locally refine the right half of the original B-splines whereas the left half remains coarse. We replace the coarse basis function which are located at the right (marked by a dashed line) by fine basis function (marked by a solid line), and in this way we create the basis shown at the bottom.

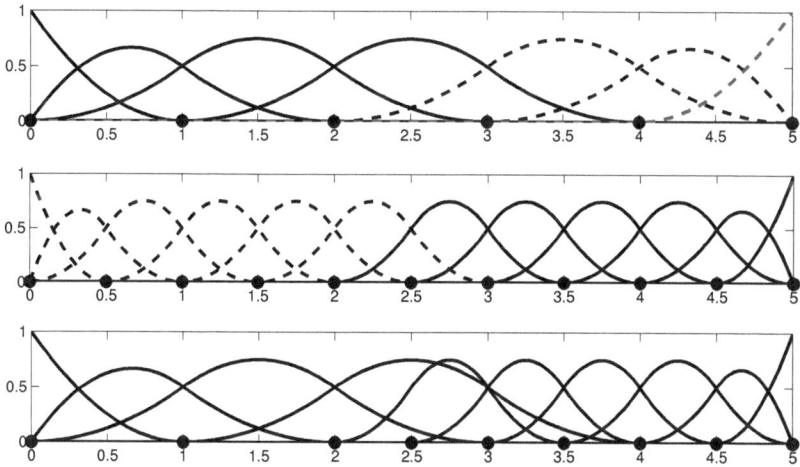

Figure 5.3: Linear independent hierarchical Basis

5.2.2 Properties of Hierarchical B-Splines

We now want to show some properties of our basis functions. First of all we have to justify to call it a basis: the linear independence of the hierarchical basis functions follows immediately from Definition 5.1.

Lemma 5.4. *The functions in $\mathcal{A}$ are linearly independent.*

Proof. Let

$$0 = \sum_{\varphi \in \mathcal{A}} d_\varphi \varphi = \sum_{\varphi \in \mathcal{A} \cap \mathcal{B}^1} d_\varphi \varphi + \sum_{\varphi \in \mathcal{A} \cap \mathcal{B}^2} d_\varphi \varphi + \ldots + \sum_{\varphi \in \mathcal{A} \cap \mathcal{B}^L} d_\varphi \varphi. \qquad (5.11)$$

Due to the linear independence of $\mathcal{B}^1$ the functions in $\mathcal{A} \cap \mathcal{B}^1$ are linearly independent as well. Only these functions are non-zero on $\Omega^1 \setminus \Omega^2$ and therefore $d_\varphi = 0$ for $\varphi \in \mathcal{A} \cap \mathcal{B}^1$ due to their local independence. Analogously, for each $\ell = 2, \ldots, L$ the functions in $\mathcal{A} \cap \mathcal{B}^\ell$ are linearly independent. Except for functions already considered before, namely in $\mathcal{A} \cap \mathcal{B}^1, \ldots, \mathcal{A} \cap \mathcal{B}^{\ell-1}$, only functions in $\mathcal{A} \cap \mathcal{B}^\ell$ are non-zero on $\Omega^\ell \setminus \Omega^{\ell+1}$. This implies that $d_\varphi = 0$ for $\varphi \in \mathcal{A} \cap \mathcal{B}^\ell$ with $\ell = 2, \ldots, L$. $\qquad \square$

The following Lemma proves the nested nature of the spaces spanned by the sequence of spline bases $A^1, \ldots, A^L$ which appear in the different levels of the hierarchy.

Lemma 5.5. *Let $A^1, \ldots, A^L$ be the spline bases considered by the iterative procedure of Definition 5.1. We have*

$$\text{span } A^\ell \subseteq \text{span } A^{\ell+1}, \qquad \ell = 1, \ldots, L - 1. \qquad (5.12)$$

Proof. Any function $f \in \operatorname{span} A^\ell$ with $\ell = 1, \ldots, L - 1$, can be expressed as

$$f = \sum_{\varphi \in A^\ell} d_\varphi \varphi = \sum_{\varphi \in A^l \backslash A_\ominus^\ell} d_\varphi \varphi + \sum_{\varphi \in A_\ominus^\ell} d_\varphi \varphi. \tag{5.13}$$

Because $A^\ell \backslash A_\ominus^\ell \subseteq A^{\ell+1}$ according to Eq. (5.8a) and $\operatorname{span} A_\ominus^\ell \subseteq \operatorname{span} A^\ell$ according to Cor. 5.2, we conclude $f \in \operatorname{span} A^{\ell+1}$. $\qquad\square$

We will discuss further properties later on in combination with the hierarchical refinement.

5.3 Adaptive Hierarchical Isogeometric Analysis

In this section we will transfer the idea of hierarchical B-Splines into the context of isogeometric analysis and discuss the realization and its properties.

In general the mesh in FEM is an irregular grid. Therefore it is possible to describe even locally refined grids within the normal framework. In contrast the isogeometric mesh has a regular tensor-product structure. The hierarchical refinement approach does not try to break this structure but extends it to an hierarchy, that resembles different resolutions, but preserves the tensor product at each level. One of the key issues here is that, in contrast to other methods borrowed from CAGD, we will take into account the very essential properties right from the beginning and assured them via construction.

We now want to extent the choice of components discussed in Sec. 4.2 to include hierarchical refinement. The components of hierarchical isogeometric analysis are

- a geometric mapping $\boldsymbol{G}$,

- set of basis functions $\mathcal{A}$, the hierarchical basis, and

- an hierarchical mesh structure .

We see that, compared to the components of isogeometric analysis, we have extended the space of functions as well as the mesh structure.

The idea of hierarchical refinement in isogeometric analysis is simply to replace the standard NURBS basis by a hierarchical one. The hierarchical basis of Definition 5.1 is defined on Ω^1, possesses favorable properties by construction and is thus a promising candidate for our objective. The idea to use a hierarchical basis for refinement was also discussed in [66] for FEM without special consideration of the element structure. Hierarchical B-splines were recently also succesfully employed for three dimensional interface problems shown in [78].

5.3.1 Hierarchical Structures

In this section we want to investigate the hierarchical mesh as well as its connection to a hierarchical basis. So far, the hierarchical spline functions have mainly been described in terms of their support. In order to access them in a more FEM-like manner we make use of the element concept in isogeometric analysis introduced in Sec. 4.2.2, where we defined the isogeometric mesh $\widehat{\mathcal{T}}$. Unfortunately, we cannot globally define the elements analogously to Def. 4.1 with respect to the intersections of the support of the hierarchical basis functions. This approach does not work over several hierarchies any more and we have to make another characterization.

An important ingredient of our approach is using a hierarchy of meshes that are constructed consecutively. More specifically, we create a sequence of meshes $\widehat{\mathcal{T}}^\ell$ with index $\ell = 1, \ldots, L$ by successive global uniform h-refinement, e.g. by inserting a knot in the middle of each non-empty knot span and therefore subdividing every isogeometric element $T^\ell \in \mathcal{T}^\ell$ that are inferred from $\widehat{\mathcal{T}}^\ell$. An example for such a sequence is shown in Fig. 5.4 and will be used for illustration from now on.

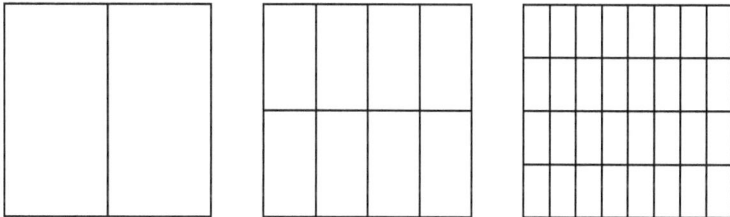

Figure 5.4: Sequence of h-refined isogeometric meshes $\widehat{\mathcal{T}}^\ell$

The mesh sequence corresponds to the levels of the spline spaces $\mathcal{S}^\ell$, and over each isogeometric mesh $\widehat{\mathcal{T}}^\ell$ a set of basis functions $\mathcal{B}^\ell$ is defined. Due to the underlying h-refinement process, this procedure can be also performed for NURBS instead of B-splines.

Hierarchical Mesh Structure

The higher smoothness offered by isogeometric analysis implies that the support of the basis functions is usually larger than in standard finite elements. For this reason, the connection between mesh and basis functions is more subtle and requires particular attention as it was discussed in Sec. 4.2.2. Now we want to extend these concepts to the hierarchical approach.

Definition 5.6. *We call a selection* $\mathcal{M} \subset \bigcup_{\ell=1}^{L} \mathcal{T}^\ell$ *a* hierarchical subdivision *if following conditions hold*

- *The interior of two arbitrary elements are disjoint,*

$$\operatorname{int} T_i \cap \operatorname{int} T_j = \emptyset \quad \forall T_i, T_j \in \mathcal{M}, T_i \neq T_j. \tag{5.14}$$

- *The union of all elements forms the whole parametric domain,*

$$\bigcup_{T \in \mathcal{M}} T = \overline{\Omega}_0. \tag{5.15}$$

An element $T \in \mathcal{M}$ is called an active element. The set of active elements with level ℓ is denoted by $M^\ell := \mathcal{M} \cap \mathcal{T}^\ell$.

Any hierarchical subdivision is a subdivision (see Def. 3.7) by definition. Moreover, it should be noted that any element $T \in \mathcal{T}^\ell$ for any level ℓ has nonempty interior by definition. Furthermore, if we select all elements from a level to be active this is a hierarchical subdivision that resembles uniform h-refinement.

An example for a hierarchical subdivision is shown in Fig. 5.5 where eight elements are selected to be active out of the three levels shown in Fig. 5.4.

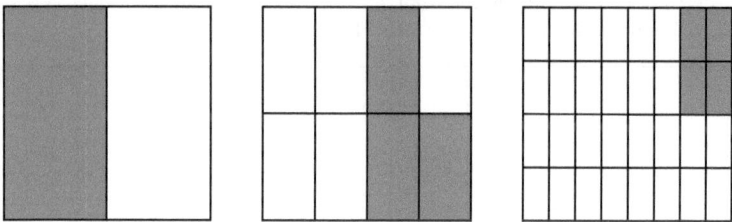

Figure 5.5: Sequence of h-refined parameter spaces with the hierarchical subdivision marked in grey

In order to establish the relation to the support of the basis functions we have to study the region filled out by particular elements in a set. For ease of notation we define the domain $U_X \subseteq \Omega_0$ to be

$$U_X := \bigcup_{C \in X} C \qquad (5.16)$$

for a set $X \subset \mathfrak{P}(\Omega_0)$.

Hierarchical Basis Functions

After having defined the active elements in the hierarchical structure we can now address the basis functions. Analogously we have to specify an appropriate selection within the hierarchy.

Definition 5.7. *The set of active basis functions $\mathcal{A}$ is defined as follows: a function $\varphi \in \mathcal{B}^k$ of level k is an element of $\mathcal{A}$ if*

$$\operatorname{supp}\varphi \subseteq \bigcup_{\ell=k}^{L} U_{M^\ell} \qquad (5.17a)$$

and

$$\operatorname{supp}\varphi \nsubseteq \bigcup_{\ell=k+1}^{L} U_{M^\ell}. \qquad (5.17b)$$

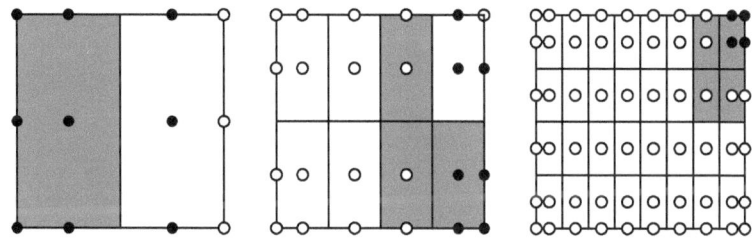

Figure 5.6: Active functions (black circles), non active functions (white circles) for degree $p_u = p_v = 2$ based on active element (grey elements)

An example is given in Fig. 5.6, which is derived from Fig. 5.5 and extended by circles to mark the Greville points (see Sec. 4.2.4).
We can identify the sequence of nested domains in Eq. (5.2) by

$$\Omega^k = \bigcup_{\ell=k}^{L} U_{M^\ell}. \tag{5.18}$$

and then Def. 5.7 coincides with Cor. 5.3. Therefore the active functions are the hierarchical basis stated by Def. 5.1 and it is justified to use the same identifier $\mathcal{A}$.

It would be a mistake to assume that active basis functions and active elements of the same level only appear together. Unfortunately, their relationship is more complex. We now state a relation between active function and the underlying hierarchical mesh, that follows directly from Def. 5.7.

Corollary 5.8. *Let $\mathcal{M}$ be a hierarchical subdivision. For each active function of level k there is an active element of level k in its support.*

Proof. Let $\varphi \in \mathcal{B}^k$ be an active function. Then with Eq. (5.17a) and (5.17b)

$$\emptyset \neq \bigcup_{\ell=k}^{L} U_{M^\ell} \setminus \bigcup_{\ell=k+1}^{L} U_{M^\ell} = U_{M^k} \subseteq \operatorname{supp}\varphi. \tag{5.19}$$

$\square$

Connection between Hierarchical Structures

We can define a set of active functions or a hierarchical basis for an arbitrary hierarchical subdivision. Corollary 5.8 shows that for the occurrence of refined basis functions refined elements are necessary. Unfortunately, this is not sufficient and it does not hold that a refined basis function occurs if we use refined elements. We will show this for a simple one dimensional example. In Fig. 5.7 three different settings for two hierarchical levels are shown. In each case a different size of the refinement region Ω^ℓ is chosen. If this region is too small like in Fig. 5.7a the support of none finer basis function is included in Ω^ℓ. This means $A_\oplus^{\ell+1} = \emptyset$ in the notation of Def. 5.1 . If the region is big enough such that the support of a finer functions fit in it, but it is not as big as the support of a coarse function, like in Fig. 5.7b, none of the coarse function is taken out of the set of active functions. This means that $A_\ominus^\ell = \emptyset$. Only in the case of Fig. 5.7c fine functions replace coarse functions.

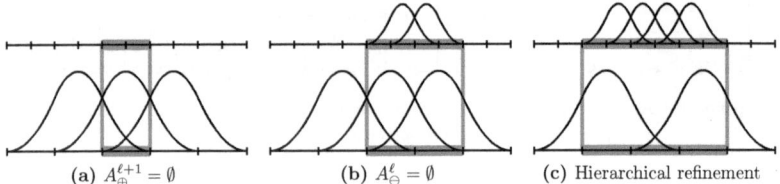

(a) $A_\oplus^{\ell+1} = \emptyset$ (b) $A_\ominus^\ell = \emptyset$ (c) Hierarchical refinement

Figure 5.7: Influence of the size of the refinement region

With these cases in mind we state following definition:

Definition 5.9. (Support Conditions)
Let $\mathcal{M}$ be a hierarchical subdivision. If for an active element $\overline{T} \in \mathcal{M}$ with level $k > 1$, i.e., $\overline{T} \in M^k$, there exists a set of active elements

$$S \subseteq \bigcup_{\ell=k}^{L} M^\ell \tag{5.20}$$

which contains the first element itself,

$$\overline{T} \in S, \tag{5.21}$$

and its union equals the union of $(p_u + 1) \times (p_v + 1)$ block of subsequent knot domains of the same level k

$$U_S = \bigcup_{r,s=0}^{p} \widehat{T}_{i+r,j+s}^k \tag{5.22}$$

with indices i and j only dependent on $\overline{T}$, the hierarchical subdivision $\mathcal{M}$ fulfills the weak support condition.
If the set S fulfills Eq. (5.21) and its union equals the union of $(p_u + 1) \times (p_v + 1)$ block of subsequent knot domains of the lower level $k - 1$

$$U_S = \bigcup_{r,s=0}^{p} \widehat{T}_{i+r,j+s}^{k-1} \tag{5.23}$$

with indices i and j only dependent on $\overline{T}$, the hierarchical subdivision $\mathcal{M}$ fulfills the strong support condition.

Due to the fact that a knot domain does not increase with raising levels it directly follows that a hierarchical subdivision that fulfills the strong support condition also fulfills the weak support condition.

For an element from level $k > 1$, the set S can be viewed as the minimal refinement region that has to be filled out with knot domains of level k or higher. The weak support condition demands that S has the size of the support of a function of level k such that it is active. This ensures that grid refinement takes effect on the basis functions. Otherwise, it would be impossible to replace the coarse function by finer ones. Figure 5.8 shows an example with $p_u = p_v = 2$, which we will use to illustrate this concept. The initial subdivision is shown in Fig. 5.8a and does not fulfill the weak and therefore also not the strong support condition. For example, we cannot find a set S that fulfills Eq. 5.22 for the marked element. This is fixed in Fig. 5.8b where an additional element is refined and a possible set S is marked in grey. Of course, in order to fulfill the weak support condition we have to consider all elements in the subdivison. In particular for the element of higher level, for example the one marked in Fig. 5.8c, a suitable S marked in grey is found.

For the strong support condition the minimal refinement region S has to be as large as the support of a coarse basis function from the previous level $(k-1)$, which is the block of $(p+1) \times (q+1)$ knot domains of level $(k-1)$. In this case this function is replaced by the finer ones. In Fig. 5.8 the first three subdivision all do not fulfill the strong support condition. Only if we further refine, we get the subdivision shown in Fig. 5.8d and we are able to identify the set S marked in grey that fulfills Eq. 5.23 for the marked element. Still, we again have to consider all elements and therefore we also have to refine the element of level two up to level three. The result is shown in Fig. 5.8e, which finally fulfills the strong support condiditon.

We can now establish the connection between elements and basis functions in conjunction with Cor. 5.8.

Lemma 5.10. *Let $\mathcal{M}$ be a hierarchical subdivision that fulfills the weak support condition. For each element T^k at level k there exists at least one active basis function of level k that is non-zero over the element.*

Proof. According to the weak support condition there exists a set of active elements $S \subseteq \bigcup_{\ell=k}^{L} M^\ell$, such that $U_S = \bigcup_{r,s=0}^{p} \hat{T}^k_{i+r,j+s}$. This is the support of the basis function B^k_{ij} according to Cor. 4.3. Then it holds that supp $B^k_{ij} = U_S \subseteq \bigcup_{\ell=k}^{L} U_{M^\ell}$ and supp $B^k_{ij} \not\subseteq \bigcup_{\ell=k+1}^{L} U_{M^\ell}$, because $T^k \in S$. With Def. 5.7 it follows that this basis function is active. $\square$

In the following we will always abide the strong support conditions out of two reasons: first, we will see in Sec. 5.3.2 that we may only fulfill the partition of unity property with it. Second, in combination with refinement criteria and error estimation discussed in Sec. 5.3.4 we can easily enforce the strong support condition. This is not the case if we would only want to enforce the weak support condition alone.

5.3.2 Hierarchical Refinement

The concepts of active elements and active basis functions put us in a position to develop a refinement procedure. Starting from a normal standard tensor-product surface defined in the space $\mathcal{S}^0$, the elements to be refined are replaced by elements of a higher level. More specifically, as we employ uniform h-refinement a single element is subdivided

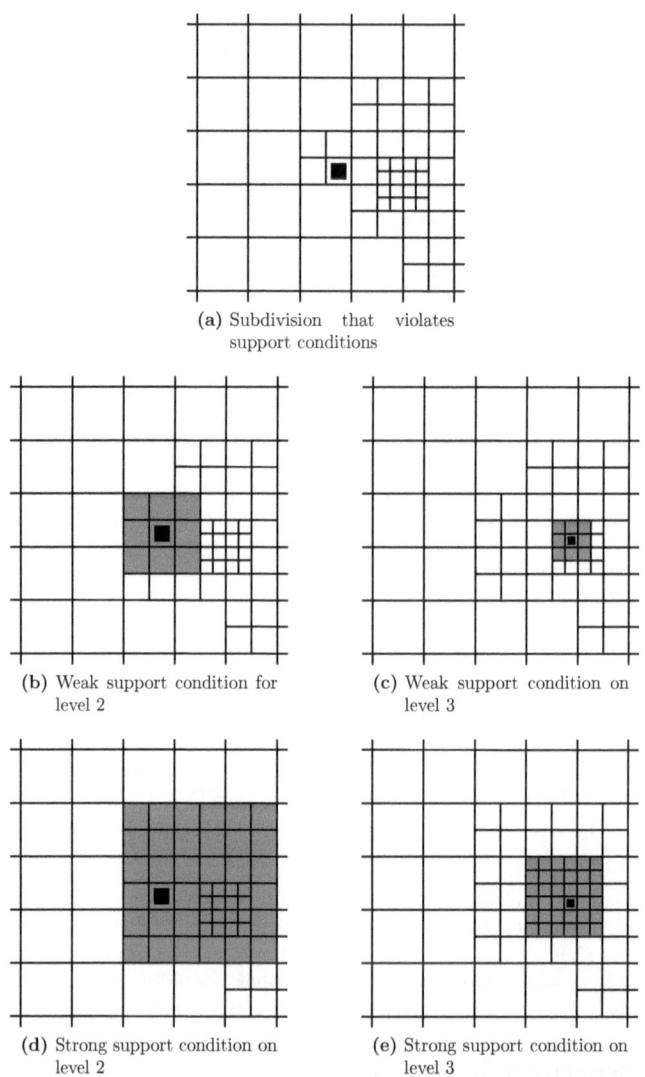

(a) Subdivision that violates support conditions

(b) Weak support condition for level 2

(c) Weak support condition on level 3

(d) Strong support condition on level 2

(e) Strong support condition on level 3

Figure 5.8: Example for support conditions: the active element $\overline{T}$ is marked black and the corresponding set of active elements S is marked in grey.

into four smaller elements. The underlying spline spaces $\mathcal{S}^\ell$ therefore are created by inserting knots that splits each non-empty knot interval into halves. Again, the small element can be refined as well, so that we get a refined grid over several levels. If we continue to subdivide other active elements, Lem. 5.11 ensures that this procedure always enlarges the current space and we get a sequence of nested spaces.

In order to simplify the graphical representation of the hierarchies, we visualize the refinement by showing all active elements within the same domain. For degrees $p_u = p_v = 1$ and $p_u = p_v = 2$ this is done in Fig. 5.9. Again, we use the configuration of Fig. 5.6 and represent the active basis functions by Greville points. The same approach will be used to visualize the refinement process for the computational examples in Sec. 5.5.

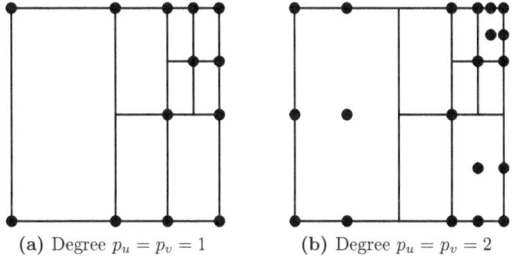

(a) Degree $p_u = p_v = 1$ (b) Degree $p_u = p_v = 2$

Figure 5.9: Refined grid with Greville points of the active functions

In the refinement process, the strong support condition must be taken into account so that the marked region is large enough. This is the only point in our approach that needs extra attention, and how to take care of it, is discussed in the context of refinement criteria and error estimation in Sec. 5.3.4. The selection of elements at different levels created in this way contains all the information that we need to compute the corresponding basis functions.

Refinement Behaviour

From the discussion so far it is not obvious how the refinement actually behaves, particular in view of the problems T-splines may face when dealing with refinement along a diagonal. Global refinement is performed by means of knot insertion and therefore follows the direction of the knot lines. From an adaptive refinement procedure we would expect that it also can handle regions that are not aligned with knot lines.

As an example we look at a square domain $[0, 1]^2$ with the identity as its parameterization and aim at refining the diagonal for degree $p_u = p_v = 2$. After two successive h-refinements in every parameter direction we obtain 16 elements in total in the starting grid. Next, the elements along the diagonal are marked for refinement, taking the strong support condition into account. That means we refine $(p_u + 1) \times (p_v + 1)$-blocks into smaller ones. In Fig. 5.10 the resulting grid is displayed. It can be seen that a high resolution at the diagonal can be achieved, nevertheless with extra layers or transition regions between the coarsest and the finest level. This moderate change of the resolution is not enforced but an effect of the refinement itself. Obviously this is a spatially limited effect because it only affects the support of changed basis functions, and accordingly, the refinement will not propagate away from the region of interest. Nevertheless, for the hierarchical refinement to take effect, the initial grid has at least so many elements that the support condition allows local refinement. This is the reason, why we have chosen a starting grid with 4×4 element.

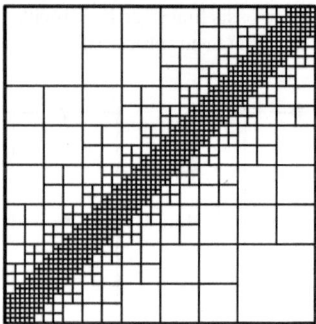

Figure 5.10: Refinement along the diagonal for degree 2

Some approaches to local refinement decrease the continuity, typically by insertion of multiple knots, to limit the propagation of the refinement. In our case of hierarchical refinement this is not necessary and the discretization preserves the initial degree of continuity. Another insightful test example in this context is the advection-diffusion problem discussed in Sec. 5.5.

Properties as a Refinement Method

Now, we want to summarize and discuss the properties of hierarchical refinement from the point of view introduced in Sec. 5.1.1. Some of the proofs have already been shown in Sec. 5.2.2.

One of the most important properties is that the mesh refinement does not propagate along the knot lines like for standard B-spline or T-spline refinement. Hierarchical refinement is **local** in the sense that the refinement only takes place at the elements marked for refinement. There is no correction step afterwards that may trigger further refinement steps. The only condition is that the refined region has to have a certain size so that it fulfills the support condition (see Def 5.9). It is ensured that the active basis functions are **linearly independent**, proven by Lem. 5.4, and therefore form a basis. The following Lemma ensures that the spline spaces are **nested**.

Lemma 5.11. *Consider the two sequence of nested domains*

$$\Omega^L \subseteq \Omega^{L-1} \subseteq \dots \subseteq \Omega^1 \quad and \quad \widehat{\Omega}^L \subseteq \widehat{\Omega}^{L-1} \subseteq \dots \subseteq \widehat{\Omega}^1, \tag{5.24}$$

together with the corresponding hierarchical bases A^ℓ and $\widehat{A}^\ell$, constructed according to Definition 5.1, for $\ell = 1, \ldots, L$. If $\Omega^\ell \subseteq \widehat{\Omega}^\ell$, for $\ell = 1, \ldots, L$, then

$$\operatorname{span} A^\ell \subseteq \operatorname{span} \widehat{A}^\ell. \tag{5.25}$$

Proof. Since $\Omega^\ell \subseteq \widehat{\Omega}^\ell$, for $\ell = 1, \ldots, L$, the sequence $\{\widehat{\Omega}^\ell\}_{\ell=1,\ldots,L}$ covers in each level a wider area of the domain by refined functions of the next basis in the underlying sequence $\{\mathcal{B}^\ell\}_{\ell=1,\ldots,L}$ and, in particular, $\operatorname{span} A^1 \subseteq \operatorname{span} \widehat{A}^1$. If we use the notation in Eq. (5.8a) we get $A_\oplus^{\ell+1} \subseteq \widehat{A}_\oplus^{\ell+1}$ and $A_\ominus^\ell \subseteq \widehat{A}_\ominus^\ell$ for $\ell = 1, \ldots, L - 1$. Furthermore we have $\operatorname{span}(\widehat{A}_\ominus^\ell \backslash A_\ominus^\ell) \subseteq \operatorname{span} \widehat{A}_\oplus^{\ell+1}$ due to Cor. 5.2 for $\ell = 1, \ldots, L - 1$. Hence, the space spanned by the basis A^ℓ is contained in the space spanned by $\widehat{A}^\ell$ for $\ell = 1, \ldots L$. $\square$

The hierarchical basis does not fulfill the **partition of unity property** on its own. If the strong support condition is fulfilled this can be fixed.

Lemma 5.12. *Let $\mathcal{A}$ be the hierarchical basis to the hierarchical subdivision $\mathcal{M}$ that fulfills the strong support condition. Then there exists a scaled basis $\mathcal{W} := \{\tau_\varphi \varphi : \varphi \in \mathcal{A}\}$ with given $\tau_\varphi > 0$ such that $\mathcal{W}$ forms a partition of unity*

$$\sum_{\varphi \in \mathcal{A}} \tau_\varphi \varphi = 1. \tag{5.26}$$

Proof. The constant function 1 belongs to the initial spline space and there also to the span A^0. Lemma 5.5 ensures that it belongs to all subsequent A^ℓ and especially $\mathcal{A}$. Therefore we have the representation

$$1 = \sum_{\varphi \in \mathcal{A}} \tau_\varphi \varphi. \tag{5.27}$$

If $\tau_\varphi \neq 0$ for all $\varphi \in \mathcal{A}$ this is a basis that fulfill the partition of unity property. We refer to the proof of Lem. 5 in [95] and the discussion of it, which shows that $\tau_\varphi > 0$ for any $\varphi \in \mathcal{A}$ if the strong support condition is fulfilled. $\square$

If the hierarchical subdivision consists only of element from the same level just get the usual uniform refinement. Therefore hierarchical refinement has **uniform refinement as a special case**. The **representability of the geometry** is assured by Lem. 5.5 and therefore it is still possible to represent the geometry exactly in the refined space. Due to the use of several hierarchies hierarchical refinement does not have a unified representation, but we will see in Sec. 5.4 that this is not a crucial problem. Built upon an initial arbitrary spline space this refinement can be applied to any degree and any continuity and therefore it is **suitable for any spline space**.

All in all the essential properties needed for using the basis functions in isogeometric analysis are ensured by construction.

5.3.3 Hierarchical Reference Element

Although it is convenient to look at a hierarchical subdivision and its active basis functions as a whole, the component that describes the basis functions from an element point of view is still missing. In FEM this was done in a straightforward way by the reference element (see Sec. 3.3.3) and in standard isogeometric analysis by the isogeometric reference element (see Sec. 4.4.1). In contrast for each member of an hierarchical mesh we cannot use the isogeometric reference element alone. Due to the bigger support of the non-refined functions there are more non-zero basis functions than in the original B-spline spaces and the function may belong of different levels.

Multiple Isogeometric Reference Elements

In order to cope with all of this, we have to extend the isogeometric reference element that only holds information about one level. Therefore we define the hierarchical isogeometric reference element of level k as a sequence of reference elements from level $\ell = 1, \ldots, k$. Figure 5.11 visualizes this for a level 3 element. We see the hierarchical subdivision with an element marked in green and its two parent elements of lower level marked in red and blue in Fig. 5.11a. The hierarchical reference element consists of the three corresponding reference elements as seen in Fig. 5.11b, whereas each of the reference elements resembles its level. It should be remarked that in the hierarchical

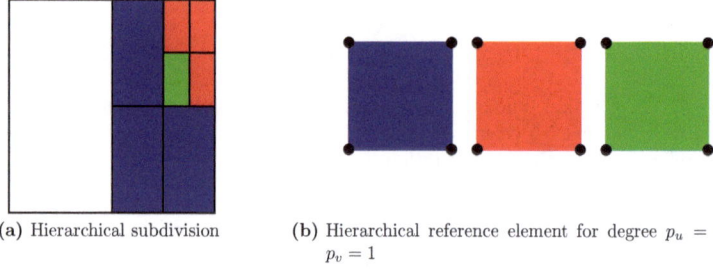

(a) Hierarchical subdivision   (b) Hierarchical reference element for degree $p_u = p_v = 1$

Figure 5.11: Hierarchical isogeometric reference element

subdivision these elements have different size, but this has no influence on the hierarchical reference element. It only indicates for all level, which basis functions are non-zero on this element, just like for the non-hierarchical case.

Active Functions on Hierarchical Elements

Although the number of basis functions on each level of the hierarchical reference element is still constant, the number of active functions is not. Each element has its own selection of active functions over the hierarchical reference element. The initial condition is the configuration where all elements at the lowest level as well all their basis functions are active. If we now recursively add a level $\ell + 1$, just like in the recursive definition of the hierarchical basis (see Def. 5.1) we need to add finer functions that are members of $A_{\oplus}^{\ell+1}$ and remove coarse functions that belong to $A_{\ominus}^{\ell}$. This is determined according to Def. 5.7 through the support extension of the element.

We illustrate the active basis functions over the reference element by an example. For the sake of simplicity we look at a hierarchical basis of degree one. In Figure 5.12a the same the hierarchical subdivision like in Fig. 5.11a is shown. We want to describe the configuration of the element marked in grey. The active basis function are symbolized by dots, whereas those that are nonzero on the grey element are colored according to their level. Blue stands for the first, red for the second and green for the third level, just like the elements in Fig 5.11. We will also use this color scheme again in Sec. 5.5 when presenting some numerical examples. Finally, the corresponding reference elements are shown in Fig. 5.12b. Again we color the active basis function according to the scheme above and express the information about which element are active over our chosen element by this.

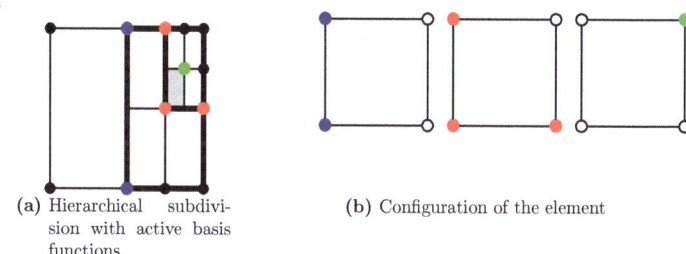

(a) Hierarchical   subdivision with active basis functions

(b) Configuration of the element

Figure 5.12: Active basis functions on hierarchical reference element

These concepts are essential for the implementation of hierarchical refinement and will be discussed further in Sec. 5.4.2.

5.3.4 Refinement Criterium and A Posteriori Error Estimation

After we have discussed how to create a locally refined hierarchical subdivision in Sec. 5.3.2 we need to investigate the criterium for marking elements for refinement. Just like for FEM we may use an element-wise error estimator for this task.

Error Estimation in Isogeometric Analysis

Error estimation was already introduced for FEM in Sec. 3.4.4, but for isogeometric analysis not a lot is known. It suggests itself to adapt the ideas from FEM to isogeometric analysis again.

Unfortunately, it is not straightforward to transfer, for example, residual error estimators to isogeometric analysis. For evaluation of the residual the contained differential operators have to be transformed to parametric space, which may be much more involving than for the bilinear form of the variational problem only. For example, for the Poisson problem we need the Laplace operator instead of gradients only.

Averaging error estimators or more general recovery error estimators can be applied on hierarchical subdivision, but their implementation is sophisticated. In contrast to earlier we need to fetch neigboring information, that may also be very irregular due to the locally refined structure of the subdivision. Both approaches were discussed in [49], where also further details can be found.

In the following we will focus on the multilevel error estimator. As it was discussed before it is based on adding additional bubble function for each element. This can be easily transfered into isogeometric analysis Using the hierarchical subdivision as introduced in Sec. 5.3.1, we can define the bubble functions W_k on their preimage and get as an augmenting subspace

$$\mathcal{W}_h = \text{span}\left\{W_k \circ \boldsymbol{G}^{-1}, \ k = 1, \ldots, n_{el}\right\}, \tag{5.28}$$

with n_{el} is the number of elements, i.e., $n_{el} = |\mathcal{M}|$. We use this subspace for a multilevel error estimator as described in Sec. 3.4.4 and therefore solve the modified varational problem for the error and compute additional submatrices. This results in an estimation of the error on each element, and by means of a marking criterion a set of elements can be selected.

Mesh Adaption

The marking criteria either provided by an error estimator or another source of information is used to mark element for refinement. However, this procedure has to be adapted to the hierarchical refinement since it is not always guaranteed that the refined grid does fulfill a support condition. One possibility to ensure this condition is a correction step after the elements have been marked, but there is another, easier way. The strong support condition (see. Def. 5.9) demands that not only single elements but complete supports of the coarse basis functions are marked for refinement. A simple approach to ensure this is to transfer the element-wise error indicator to the basis functions by a modified error indicator $\tilde{\eta} : \Phi \to \mathbb{R}_0^+$ that forms the average

$$\tilde{\eta}(\varphi) = \frac{1}{|S(\varphi)|} \sum_{T \in S(\varphi)} \eta(T) \tag{5.29}$$

with $S(\varphi) := \{T \in \mathcal{A} \ : \ T \subset \text{supp}\,\varphi\}$. Analogously to a marking criteria for elements, we select the functions to be refined based on the data provided by $\tilde{\eta}$. In the mesh, only those elements that are in the union of the supports of the marked basis functions are refined. In this way the strong support condition Eq. (5.23) is satisfied by construction. In Figure 5.13 these two steps of the marking criteria are visualized for a half-disk geometry. The element-wise error estimate is shown in Fig. 5.13a. According to Eq. 5.29 the estimator for each basis function is computed. As usual we use Greville points to visualize them and a stem diagram for the new refinement criteria is shown in Fig 5.13b.

5.4 Implementation Issues

Just as it was done before we want to look at some algorithmic and implementation aspects. This section is an extension of Sec. 4.6, because the uniform refinement is only a special case of the hierarchical refinement in isogeometric analysis.

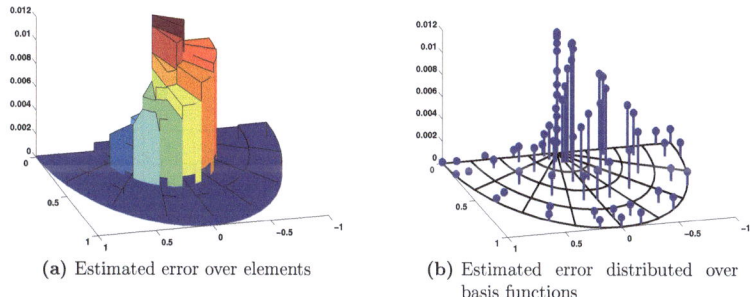

(a) Estimated error over elements

(b) Estimated error distributed over basis functions

Figure 5.13: Distribution of estimated error

5.4.1 Data structures

The hierarchical structures introduced in Sec. 5.3.1 need to be resembled in the data structures. In principle, we keep the data structures described in Sec. 4.6.1 for each level. There are several aspects we have to deal with:

- the hierarchical subdivision (or the distribution of the active element) over the hierarchy of isogeometric meshes $\widehat{\mathcal{T}}^\ell$,

- the distribution of active basis functions over the hierarchy of functions sets $\mathcal{B}^\ell$,

- the relation between these two previous points through hierarchical elements.

Each element is characterized by its level ℓ and its number within the level. An example shown in Fig. 5.14a. The hierarchical subdivision is described by a list of elements. Table 5.3a shows these for the example introduced before. Each function is uniquely identified by it level and its enumeration within the level. In Fig. 5.14b the numbers of the active functions are shown. All basis functions are gathered in an element list shown in Tab. 5.3b. The relation between elements and basis functions is described by the hierarchical reference elements (see Sec. 5.3.3) and their local enumeration of the basis functions at each level. These can be inferred from of the described data and stored separately.

5.4.2 Hierarchical Refinement

Until now it was not necessary to discuss the implementation of refinement operations, because they were always introduced as CAD algorithms before (see Sec. 2.4.2). For hierarchical refinement this is not the case and therefore we want to to investigate them in this section.

We assume that we have a hierarchical subdivision up to level ℓ and now want to add a refinement region $\Omega^{\ell+1} \subseteq \Omega^\ell$. Therefore it only consists out of marked element of the level ℓ. The element list itself is modified by adding the child elements and removing the marked elements.

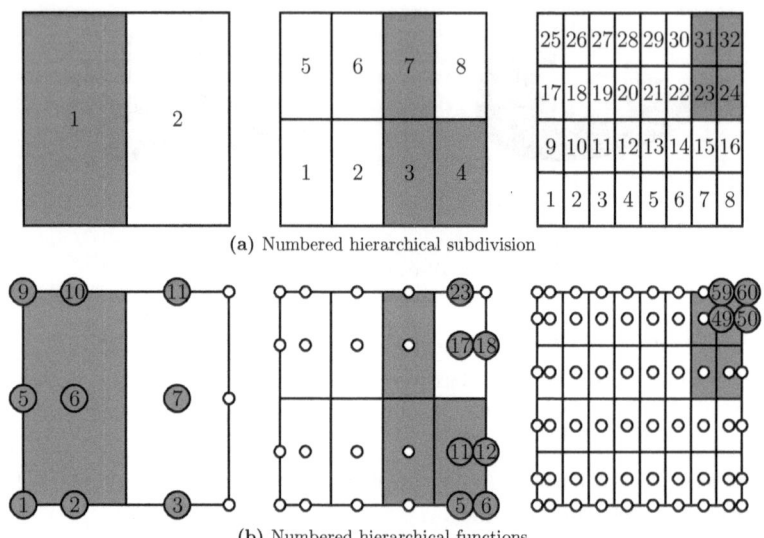

(a) Numbered hierarchical subdivision

(b) Numbered hierarchical functions

Figure 5.14: Example for enumeration in the Hierarchy

Table 5.3: Data structures for the hierarchical subdivision

(a) Elements

number	1	3	4	7	23	24	31	32
level	1	2	2	2	3	3	3	3

(b) Basis function

number	1	2	3	5	6	7	9	10	11	5
level	1	1	1	1	1	1	1	1	1	2

number	6	11	12	17	18	23	49	50	59	60
level	2	2	2	2	2	2	3	3	3	3

Now we need to compute, which functions are set active by the hierarchical subdivision. We start by computing the configuration of the reference element for each marked element (see. Sec. 5.3.3. In order to do so the support extension of the element at its hierarchy level is extracted and through other marked elements at the same level some basis functions of level ℓ may be deactivated (see Eq. (5.8b)). The refined child elements inherit this configuration for level ℓ and below from their marked parent elements. Similarly this is done for the level $\ell + 1$ for each refined element. Which of the new added functions at level $\ell + 1$ are set active is determined by Eq. (5.8c) applied to the support extension of the element, again at the same level $\ell + 1$. Note that for the last two operations only information about element of the same level are needed, which are easily obtained. It is not necessary to switch between the levels of the hierarchies for this. The whole algorithm is summarized in Alg. 5.1.

Algorithm 5.1 Level refinement algorithm

> **for all** marked elements **do**
>> set child elements active
>> deactivate parent element
> **end for**
> refresh element list
> **for all** marked elements T_m **do**
>> update element configuration for level ℓ
>> **for all** child elements of T_m **do**
>>> store reference element configuration for levels less or equal ℓ
>> **end for**
> **end for**
> **for all** new elements **do**
>> update element configuration for level $\ell + 1$
>> store reference element configuration for current levels
> **end for**
> refresh global basis list

The general refinement algorithms consists out of the iterative usage of Alg. 5.1 on the levels $\ell = 1, \ldots, L$, whereas the marked element are from level ℓ, respectively. This constructs the hierarchical basis just like in Def. 5.1 as well as the configuration of the element.

5.4.3 Matrix Assembly

The assembly of the system matrices as already discussed for FEM in Sec. 3.5.2 and isogeometric analysis in Sec. 4.6.3 benefits enormously from the mesh structure. The fact that we have a subdivision of the computational domains that contains the intersections of the basis functions is lead back to their use for this purpose. Therefore the matrix assembly for a hierarchical refined grid is very similar to the assembly described in Sec. 4.6.3 with the help of the data structures described in Sec. 5.4.1. We still can use a loop over each element in the element list, although they may be from different levels. For each element we use the active local basis functions given by the configuration of the element to integrate them within the bilinear form. Simple evaluations of a basis function is the same as in standard isogeometric analysis. The same routines are only employed for each hierarchical level. So only the non-zero basis functions on a specific level may be computed at once. The evaluation of the geometry is obtained through one instance within the hierarchy, typically the lowest one. All in all we can keep the assembly algorithm in Alg. 4.1.

Fortunately, due to the hierarchical structure we also have the possibility to use preevaluation. It turns out to be more complex than in the non-hierarchical case. Again we want to evaluate the univariate B-splines before the element loop. Each isogeometric mesh defines its own quadrature points and it does not suffice to evaluate the basis functions of the same level only. Instead on an isogeometric mesh of level ℓ all basis functions of level less or equal ℓ need to be evaluated at these points. This is related to the hierarchical reference element of level ℓ, which also holds all basis functions of less or equal ℓ. So, in the hierarchical case we need to invest some more evaluations in the beginning, but it is still worthwhile and speeds up the assembly significantly.

5.4.4 Treatment of Boundary Conditions

The treatment of boundary condition is very similar to uniform refined grids (see Sec. 4.6.4). The only difference is the fact that the degrees of freedom and the segments of the boundary curve are distributed over several hierarchies.

For Dirichlet boundary conditions you have to choose the right values for the coefficients again and e.g. a spline fit has to be conducted to obtain these values. In contrast to the uniform refined case we do not have direct access to all active functions at the boundary as well as their global number. Instead we use the regular tensor product structure at each level to get the active element at the boundary. An example is shown in Fig. 5.15 where the elements at the upper boundary are selected. Through local enumeration of the basis functions at each element we can get the boundary functions as well as their global number.

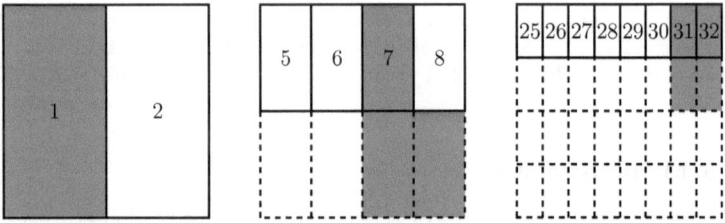

Figure 5.15: The upper boundary element are selected from each level (continuous line) whereas only the active elements are used (marked in gray)

We want to point out here that it may occur that even if the Dirichlet boundary function is constant, we can not just prescribe the constant value, because the hierarchical basis does not fulfill the partition of unity property. Alternatively, the scaling to get a partition of unity has to be computed beforehand (see Lem. 5.12).

For Neumann boundary conditions we need to fetch the pieces of the boundary curve, because we do not have a global knot vector available. Fortunately, due to the tensor product structure elements at the boundary at each level are known just like in the Dirichlet case. But instead of looking at the local basis functions, each corresponding face of each active boundary element contributes to the overall boundary integral.

5.4.5 Visualization

The visualization is again similar to the non-hierarchical case described in Sec. 4.6.5. If we set all coefficients that are not used in the hierarchy equal to zero, we can visualize each level just like in Sec. 4.6.5. The hierarchical solution is nothing more than the sum over each solution

$$u = \sum_{\ell=1}^{L} u^{\ell}. \tag{5.30}$$

The same holds if we need derivatives of the solution, e.g. for stress calculations. They are evaluated at each level by the usual algorithms and then added up to the hierarchical solution.

5.5 Numerical Simulation Cases

In this section we will present some simulation results obtained with hierarchical refinement. We will investigate several problems based on the models introduced in Sec. 3.1 and discuss their refinement and convergence behavior. Some of them were also used in Sec. 4.7 in combination with uniform refinement.

5.5.1 Advection-Diffusion

We start with an benchmark example to illustrate the locality of the hierarchical refinement. The problem consists of solving the advection-diffusion equation

$$-\kappa\Delta u + \boldsymbol{v} \cdot \nabla u = f \tag{5.31}$$

on a unit-square with discontinuous Dirichlet boundary conditions shown in Fig. 5.16 and was already introduced in [58].

The diffusion coefficient is set very small ($\kappa = 10^{-6}$) compared to the advection velocity $\boldsymbol{v} = (\sin\theta, \cos\theta)^T$. Therefore, very sharp layers arise and the equations need to be stabilized in order to obtain reasonable numerical results. We employ the SUPG stabilization with the same parameter

$$\tau_T = h_T/(2||\boldsymbol{v}||) \tag{5.32}$$

with $h_T = \mathrm{diam}(T)/(\sqrt{2}\max(\sin\theta, \cos\theta))$ as it was done in [58] and [42]. We refer to [41] for more details about stabilization techniques.

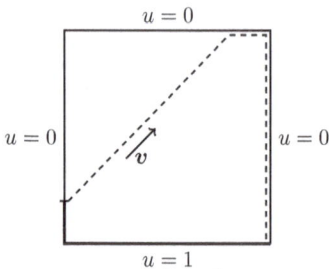

Figure 5.16: Boundary conditions for the advection diffusion problem. The expected front is shown by a dashed line for $\theta = \frac{\pi}{4}$.

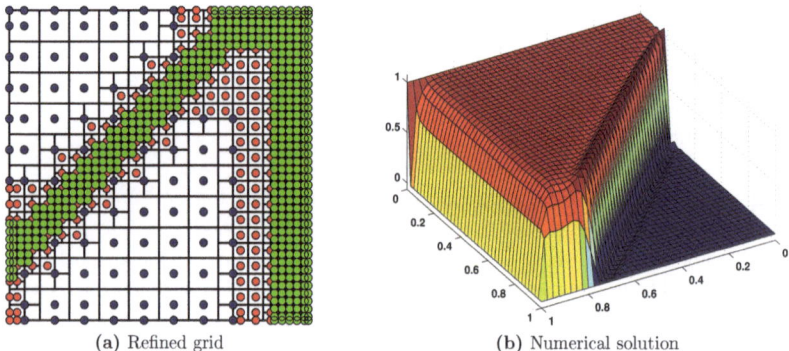

(a) Refined grid   (b) Numerical solution

Figure 5.17: Advection diffusion problem with ansatz degree 2

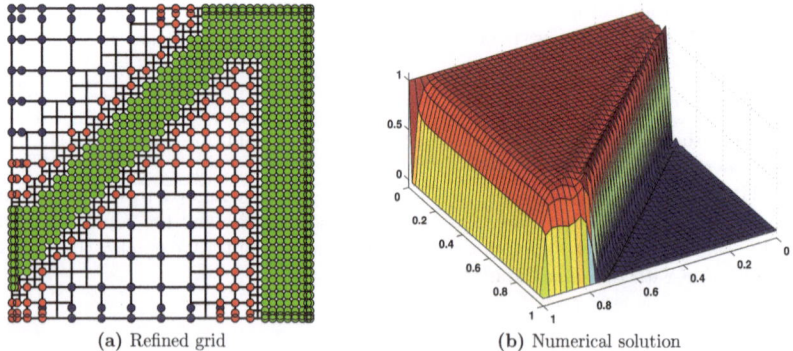

(a) Refined grid   (b) Numerical solution

Figure 5.18: Advection diffusion problem with ansatz degree 3

As a parameterization we can simply use the identity and we furthermore choose the knot vectors in such a way that we have a knot at the discontinuity of the Dirichlet boundary condition. The elements that are crossed by the sharp layer are marked to increase the resolution where it is needed most. Due to the diagonal expansion of the marked region it is interesting how the refinement behavior adapts to this.

The grid and the numerical result for degree two and three is shown in Fig. 5.17 and Fig. 5.18, respectively. We use the same color scheme blue - red -green already shown in Sec. 5.3.3 to color the different level of the Greville points or active basis functions. It is remarkable that the refined region stays near the initially marked selection. No propagation occurs like with T–Splines as shown in [42], which would lead to a global refinement. Of course, the numerical solution, which is plotted in Fig. 5.17b and 5.18b, shows the sharp layers due to the higher resolution at the appropriate locations.

5.5.2 Heat Conduction of a Half-Disk

The solution of the Poisson problem on a half-disk with adjacent Dirichlet and Neumann boundary conditions was investigated for uniform refinement in Sec. 4.7.1 and we will employ adaptive hierarchical refinement this time.

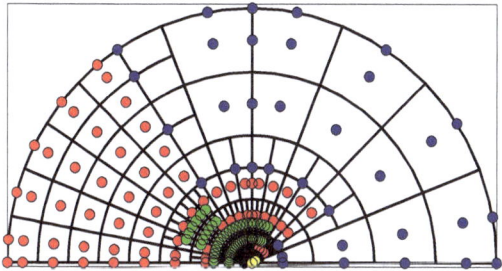

Figure 5.19: Refined grid for half-disk with 227 degrees of freedom

The geometric description includes 15 control points and we have two elements to begin with. Just like in the uniform case we apply two uniform h-refinement steps to obtain a suitable starting grid for the simulation. The multilevel error estimator introduced in Sec. 5.3.4 provides an estimate η and every element T with $\eta(T) \geq 0.75 \max_k(\eta(T_k))$ is marked for refinement. The distribution of the marking criteria was already shown in Fig. 5.13 for this example. Also the refined grid in Fig. 5.19 shows that the region near the origin, where the Dirichlet and the Neumann boundaries meet is refined.

The convergence behavior is plotted in Fig. 5.20. We can see that in this case the adaptive solution is obtained with about the half of number of degrees of freedom with similar precision. Similar holds if we apply a p-refinement step before the initial h-refinement steps to increase the degree to three.

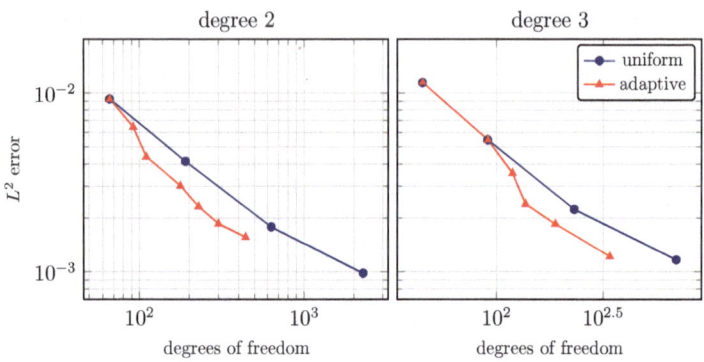

Figure 5.20: Convergence plots for the Poisson problem on the half-disk with different ansatz degrees

5.5.3 Plate with Hole

The infinite plate example was introduced in Sec. 4.7.3. Starting from the same initial parameterization of degree 2 with 28 degrees of freedom, we obtain the refined grid in Fig. 5.21. Fig. 5.22 illustrates the convergence behavior of the first principal stress component in the middle of the quarter circle with free boundary conditions.

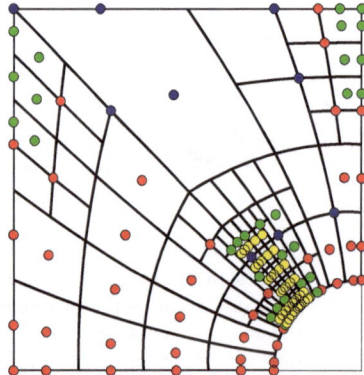

Figure 5.21: Refined grid for plate with hole problem with 110 degrees of freedom

We see that the adaptive refinement improves the convergence of this value. Unfortunately, this has no global effect in this test case, because no distinct local features are found by the error estimator. After a few refinement steps we obtain an almost uniform refined grid again. Still, the convergence of the global L^2 error is very similar for the adaptive and uniform refined cases. This example shows that in the worst

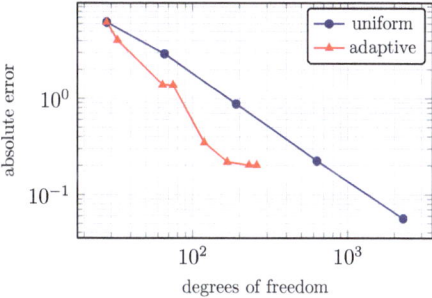

Figure 5.22: Pointwise convergence plots for the plate with hole and degree two

case of hierarchical refinement we end up with uniform refinement, which also may be suitable for some problems.

5.5.4 Circular Plate

In imitation of the three dimensional wheel example in Sec. 4.7.5 we now investigate a two dimensional circular plate with radius 1. We use the same plane stress model as in the previous example and fix the plate at the inner circular boundary with radius 0.1 and apply a force $\boldsymbol{f} = (0, 1 \cdot 10^6)^T$ at the bottom boundary, where the distance to the y-axis is less than 0.1.

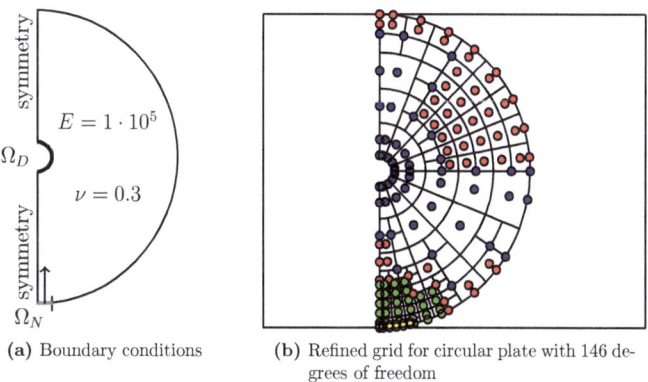

(a) Boundary conditions   (b) Refined grid for circular plate with 146 degrees of freedom

Figure 5.23: Circular plate

Due to symmetry we only simulate the right half of it with the boundary conditions and material parameters shown in Fig. 5.23a. Starting from the initial geometric description with 15 degrees of freedom and two h-refinement steps we obtain an suitable

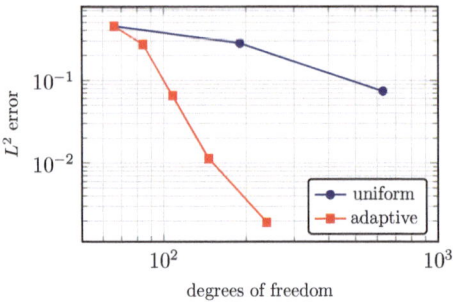

Figure 5.24: Convergence of the solution for the circular plate problem

starting grid with 66 degrees of freedom. Based on the multilevel error estimator ten percent of the elements are marked for refinement. The locally refined grid is shown in Fig. 5.23b. We see that the error estimator detects the region the force is applied, whereas in the upper part comparably less happens.

For this case we do not have an analytical solution available. Therefore we use the numerical solution under uniform refinement with 2278 degrees of freedom as the reference solution. The convergence plot is shown in Fig. 5.24 and we again observe the much higher rate of convergence of the adaptive simulation compared to the uniform one.

5.5.5 Heat Conduction on an L-shape

In this section we reuse the problem of the Laplace problem on an L-shape stated in Sec. 4.7.2, which is a good test example for local refinement. It is expected from a good error estimator that the corner singularity is detected and refinement is triggered there.

For all configuration we use the multilevel error estimator with the marking threshold $\theta = \alpha \max_k(\eta(T_k))$, where we use $\alpha = 0.5$ for the C^0 L-shape and $\alpha = 0.75$ for the C^1 L-shape. The grids for the different parameterizations, both with degree two and after four refinement steps, are shown in Fig. 5.25a and Fig. 5.25b. It is visible that the expectations were met and local refinement near the reentrant corner can be observed. Nevertheless, for the two different parameterizations the refinement differs. For the C^1 L-shape all element of the first level are marked for refinement over time in contrast to the C^0 L-shape. Furthermore some additional refinement regions at the corners can be seen, which is caused by higher deformation through the parameterization. Once again we see the influence of the parameterization on the numerical approximation, although the error estimator may be able to react to this as well.

In Fig. 5.26 the convergence plots for the L-shape for both degrees are shown. Again it can be seen that the adaptive approach performs superior to the uniform one by using lesser degrees of freedom with comparable precision. This is also the case for the convergence result for the C^1 L-shape in Figure 5.27. Nevertheless we observe that the adaptive refinement only slightly improves in the first step. This may be an

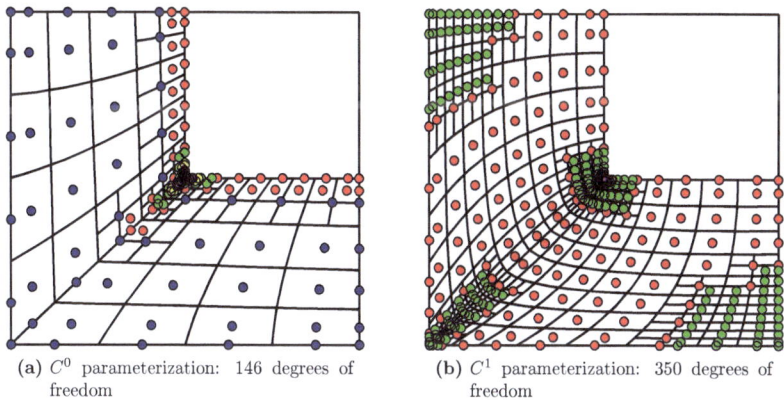

(a) C^0 parameterization: 146 degrees of freedom

(b) C^1 parameterization: 350 degrees of freedom

Figure 5.25: Refined L-shapes

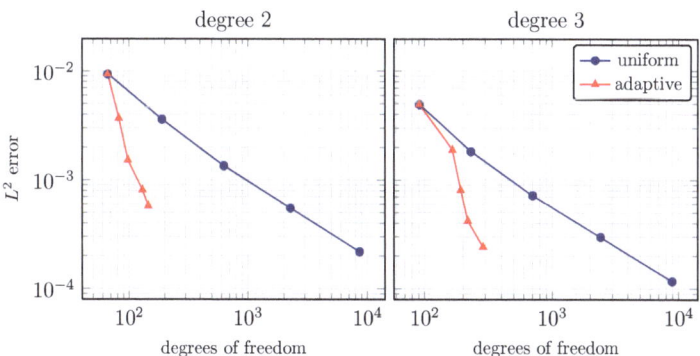

Figure 5.26: Convergence plots for the Poisson problem on the L-shape with C^1 parameterization and different ansatz degrees

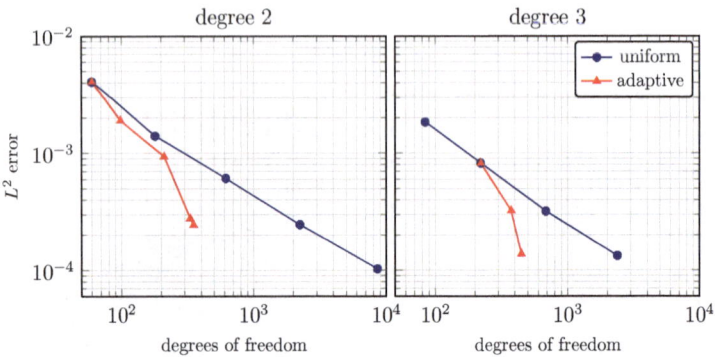

Figure 5.27: Convergence plots for the Poisson problem with C^1 parameterization and different ansatz degrees

indicator that the the initial grid is still too coarse. Therefore we started the adaptive simulation for degree three in this case after an additional h-refinement step and the result confirms this.

Properties of the Stiffness Matrix

We will now look at some properties of the stiffness matrix under hierarchical refinement. Again we employ the L-shape example and look at the condition number. The results are listed in Tab. 5.4 for the C^0 parameterization and in Tab. 5.5 for the C^1 parameterization.

Table 5.4: Condition number κ for the C^0 L-shape

(a) Degree 2

uniform		adaptive	
dof	κ	dof	κ
66	67.7	66	67.7
190	231.7	82	100.2
630	847.4	98	120.0
2278	3235.5	130	131.2
8646	12652.9	146	137.5

(b) Degree 3

uniform		adaptive	
dof	κ	dof	κ
91	113.2	91	113.2
231	361.3	165	964.4
703	1223.3	192	1376.6
2415	4457.5	219	1552.4
8911	16999.8	288	1694.0

The condition number κ for uniform refinement increases steadily just like for FEM (also cf. Sec. 4.7.3). In the adaptive case κ remains relatively low because the number of degrees of freedom does not become that high. It might still occur that it is higher than the uniform case with about the same number of degrees of freedom as seen for the C^0 L-shape and degree three in Tab. 5.4b. This may caused by the intersection of the support of basis functions from different levels and is a usual observation that you

Table 5.5: Condition number κ for the C^1 L-shape

(a) Degree 2				(b) Degree 3			
uniform		adaptive		uniform		adaptive	
dof	κ	dof	κ	dof	condition	dof	condition
60	90.2	60	90.2	220	576.0	220	576.0
180	385.9	98	135.7	684	3034.9	354	1178.2
612	2132.6	210	198.5	2380	20999.4	426	1430.7
2244	15605.6	330	797.7				
8580	119350.9	350	797.8				

make when dealing with locally refined grids. Nevertheless, the conditions numbers still behave well in this case.

We now want to have a short look at the sparse matrix structure shown in Fig. 5.28. For the L-shape with a degree two C^0 parameterization we compare the stiffness matrix for a uniform and an adaptive mesh after two refinement steps, which are shown in Fig. 5.28a and Fig. 5.28b, respectively.

We see that for the uniform refined case we get a banded matrix of the size 630×630 and nonzero entries up to the 72nd subdiagonal(Fig. 5.28c), whereas in the hierarchical refined case we get an overall smaller matrix of size 98×98, but with additional entries up to the 52nd subdiagonal due to the increased overlap between the basis functions (Fig. 5.28d). In order to minimize the influence of the enumeration on the sparsity structure we apply the reverse Cuthill-McKee algorithm on these matrices to reduce the bandwidth. The resulting sparsity structures are shown in Fig. 5.28e for the uniform with entries up to the 66th subdiagonal and in Fig. 5.28f for the adaptive case with entries up to the 35th diagonal. So, all in all we still see that due to the hierarchical refinement we get smaller matrices with higher bandwidth independently of the ordering of the degrees of freedom.

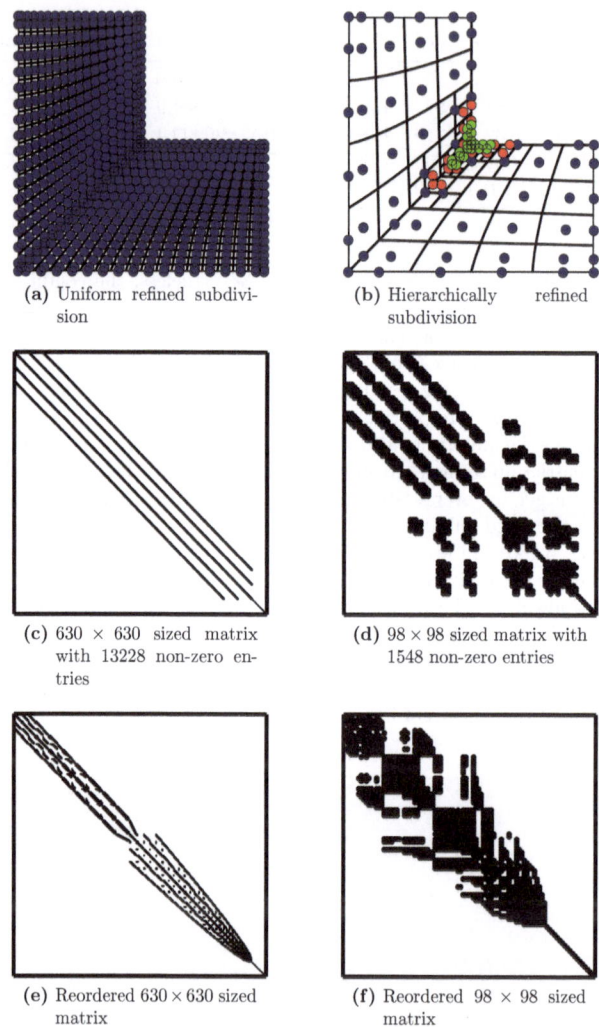

(a) Uniform refined subdivision

(b) Hierarchically refined subdivision

(c) 630 × 630 sized matrix with 13228 non-zero entries

(d) 98 × 98 sized matrix with 1548 non-zero entries

(e) Reordered 630 × 630 sized matrix

(f) Reordered 98 × 98 sized matrix

Figure 5.28: Sparse Matrix Structure for L-shape

Chapter 6

Conclusions

It is through science that we prove,
but through intuition that we discover.

(Henri Poincaré)

In this thesis we have identified and investigated the main components of isogeometric analysis and related them to well-proven concepts of FEM. Isogeometric analysis is a method for solving partial differential equations under consideration of the geometric description given by B-splines or NURBS. Although isogeometric analysis employs a global point of view and uses the basis functions given by a spline parameterization for the Galerkin projection, we were able to introduce a mesh structure that incorporates aspects from FEM as well as CAGD. Thereby we distinguish between elements and knot domains. The former are used for defining reference elements and for an efficient assembly of the system matrices and the latter are needed for evaluating B-splines. Based on this we illustrated the implementation and further techniques for increasing efficiency. Simulation examples in two and three dimensions that demonstrate the capabilities of isogeometric analysis were shown.

Since adaptivity is essential for any computational method, this thesis has also dealt with this urgent challenge in isogeometric analysis. We have introduced a local refinement approach that differs from the currently available techniques. It is based on hierarchical B-splines and it was modified such that the necessary properties, in particular linear independence, are guaranteed by construction. This local refinement approach is further supported by hierarchical structures that describe elements as well as basis functions. These structures are related to the element concept introduced for non-hierarchical isogeometric analysis, which now can be seen as a special case, and support the efficient algorithmic realization of the refinement procedure. Moreover, the connection between the mesh and the basis function is determined by the size of the refinement region compared to the size of the basis function's support. In this regard we have introduced support conditions that ensure that we deal with feasible meshes. Since this is the only restriction, the refined area remains local and we do not get global refinement by propagation like in standard isogeometric analysis. Hierarchical refinement can be applied to any degree and also works for multiple knots, which do not need to be treated in a special way. The numerical results for hierarchical refinement in combination with a multi-level error estimator have shown that it is possible to achieve the same precision like uniform refinement with less degrees of freedom and therefore emphasize the potential of this approach.

Appendix A

Software

Several implementations were done by the author during the last years and each of them has its own history and purpose. Each of them is treated in more details in the corresponding publication and just an overview is given here.

ISOGAT – Isogeometric Analysis Tutorial

This is a tutorial Matlab software that is intended to support a practical algorithmic introduction to isogeometric analysis (see [97]). Although it does not offer a wide range of possibilities it served well as a starting point for students beginning their work in isogeometric analysis.

HAIGA – Hierarchical Adaptive Isogeometric Analysis

This oject-oriented MATLAB solver packages focuses on adaptive hierarchical refinement. As the main implementation of this thesis it implements the components described in Sec. 4.6 as well as the refinement techniques described in Sec. 5.4. This software was used for computation of the examples shown in Chap. 5 and has been the implementation basis for the diploma thesis [49].

C++ Implementation

Within the EXCITING project some isogeometric solvers were implemented in C++ and contributed to the project's software toolbox. We refer the technical report [88], which describes this solver in its development stage. This solver includes the capability to solve three-dimensional problems and was used for computation of the three dimensional examples shown in Sec. 4.7 and in the diploma thesis [100]. An important library used here is the "GoTools" library by SINTEF, which implements the essential algorithms related to geometric objects. Furthermore it includes postprocessing routines e.g. for stress computation and export into the vtk file format, that can be read by Paraview (see [89]). It makes use of the Gmm++ library to store the sparse matrices and employs UMFPACK as a direct sparse solver.

Appendix B
Geometric Data

> God is in the details.
>
> *(Ludwig Mies van der Rohe)*

Half-disk

see Sec. 4.7.1 and Sec. 5.5.2

Table B.1: Half-disk

$p_u = 2, p_v = 2, \Xi_u = [0,0,0,1,1,1] \; \Xi_v = [0,0,0,0.5,0.5,1,1,1]$

No	1	2	3	4	5	6	7	8	9	10	11	12	13	14	15
C_1	-1	-0.5	0	-1	-0.5	0	0	0	0	1	0.5	0	1	0.5	0
C_2	0	0	0	1	0.5	0	1	0.5	0	1	0.5	0	1	0	0
ω	1	1	1	$\frac{1}{\sqrt{2}}$	$\frac{1}{\sqrt{2}}$	$\frac{1}{\sqrt{2}}$	1	1	1	$\frac{1}{\sqrt{2}}$	$\frac{1}{\sqrt{2}}$	$\frac{1}{\sqrt{2}}$	1	1	1

L-shape

see Sec. 4.7.2 and Sec. 5.5.5

Table B.2: C^0 L-shape

$p_u = 2, p_v = 2, \Xi_u = [0,0,0,0.5,0.5,1,1,1] \; \Xi_v = [0,0,0,1,1,1]$

No	1	2	3	4	5	6	7	8	9	10	11	12	13	14	15
C_1	-1	-1	-1	0	1	-0.6	-0.55	-0.5	0	1	0	0	0	0.5	1
C_2	1	0	-1	-1	-1	1	0	-0.5	-0.55	-0.6	1	0.5	0	0	0

Table B.3: C^1 L-shape

$p_u = 2, p_v = 2, \Xi_u = [0,0,0,0.5,1,1,1]$ $\Xi_v = [0,0,0,1,1,1]$

No	1	2	3	4	5	6	7	8	9	10	11	12
C_1	-1	-1	-1	1	-0.65	-0.7	0	1	0	0	0	1
C_2	1	-1	-1	-1	1	0	-0.7	-0.65	1	0	0	0

Plate with hole

see Sec. 4.7.3 and Sec. 5.5.3

Table B.4: Plate with hole

$p_u = 2, p_v = 2, \Xi_u = [0,0,0,0.5,0.5,1,1,1]\, \Xi_v = [0,0,0,1,1,1]$

No	1	2	3	4
C_1	-1	-1	$-\frac{1}{\sqrt{2}}$	$1-\sqrt{2}$
C_2	0	$\sqrt{2}-1$	$\frac{1}{\sqrt{2}}$	1
ω	1	$\frac{1}{2}+\frac{1}{2\sqrt{2}}$	$\frac{1}{2}+\frac{1}{2\sqrt{2}}$	$\frac{1}{2}+\frac{1}{2\sqrt{2}}$

No	5	6	7	8	9	10	11	12	13	14	15
C_1	0	-2.5	-2.5	-1.5	-0.75	0	-4	-4	-4	-2	0
C_2	1	0	0.75	1.5	2.5	2.5	0	2	4	4	4
ω	1	1	1	1	1	1	1	1	1	1	1

Circular Plate

see Sec. 5.5.4

Table B.5: Circular plate

$p_u = 2, p_v = 2, \Xi_u = [0,0,0,0.5,0.5,1,1,1]\, \Xi_v = [0,0,0,1,1,1]$

No	1	2	3	4	5	6	7	8	9	10	11	12	13	14	15
C_1	0	1	1	1	0	0	0.5	0.5	0.5	0	0	0.1	0.1	0.1	0
C_2	1	1	0	-1	-1	0.5	0.5	0	-0.5	-0.5	0.1	0.1	0	-0.1	-0.1
ω	1	$\frac{1}{\sqrt{2}}$	1	$\frac{1}{\sqrt{2}}$	1	1	$\frac{1}{\sqrt{2}}$	1	$\frac{1}{\sqrt{2}}$	1	1	$\frac{1}{\sqrt{2}}$	1	$\frac{1}{\sqrt{2}}$	1

Bibliography

[1] M. K. Agoston. *Computer Graphics and Geometric Modeling: Implementation and Algorithms*. Springer, 2005.

[2] M. K. Agoston. *Computer Graphics and Geometric Modeling: Mathematics*. Springer, 2005.

[3] M. Aigner, C. Heinrich, B. Jüttler, E. Pilgerstorfer, B. Simeon, and A.-V. Vuong. Swept volume parametrization for isogeometric analysis. In E. Hancock and R. Martin, editors, *The Mathematics of Surfaces (MoS XIII 2009)*, pages 19 – 44. Springer, 2009.

[4] M. Ainsworth and J. T. Oden. *A posteriori error estimation in finite element analysis*. John Wiley & Sons, 2000.

[5] J. Alberty, C. Carstensen, and S. A. Funken. Remarks around 50 lines of Matlab: Short finite element implementation. *Numerical Algorithms*, 20:117–137, 1999.

[6] J. Alberty, C. Carstensen, S. A. Funken, and R. Klose. Matlab implementation of the finite element method in elasticity. *Computing*, 69:239–263, 2002.

[7] K. Atkinson and W. Han. *Theoretical numerical analysis: a functional analysis framework*. Springer, 2009.

[8] F. Auricchio, L. Beirão da Veiga, A. Buffa, C. Lovadina, A. Reali, and G. Sangalli. A fully "locking-free" isogeometric approach for plane linear elasticity problems: A stream function formulation. *Computer Methods in Applied Mechanics and Engineering*, 197:160–172, 2007.

[9] I. Babuška, U. Banerjee, and J. E. Osborn. Survey of meshless and generalized finite element methods: A unified approach. *Acta Numerica*, 12:1–125, 2003.

[10] I. Babuška and W. Rheinboldt. Error estimates for adaptive finite element computations. *SIAM Journal Numerical Analysis*, 15:736–754, 1978.

[11] I. Babuška and T. Strouboulis. *The finite element method and its reliability*. Oxford University Press, 2001.

[12] W. Bangerth. Assembling matrices in deal II. Technical report, ETH Zürich, 2002.

[13] W. Bangerth, R. Hartmann, and G. Kanschat. deal.II – a general purpose object oriented finite element library. *ACM Transactions on Mathematical Software*, 33:24/1–24/27, 2007.

[14] W. Bangerth and G. Kanschat. Concepts for object-oriented finite element software – the `deal.II` library. Preprint 99-43 (SFB 359), IWR Heidelberg, 1999.

[15] W. Bangerth and O. Kayser-Herold. Data structures and requirements for hp finite element software. *ACM Transactions on Mathematical Software*, 36:4/1–4/31, 2009.

[16] R. E. Bank and R. Smith. A posteriori error estimates based on hierarchical bases. *SIAM Journal Numerical Analysis*, 30:921–935, 1993.

[17] S. Bartels, C. Carstensen, and A. Hecht. P2Q2Iso2D=2D isoparametric FEM in Matlab. *Journal of Computational and Applied Mathematics*, 192:219–250, 2006.

[18] Y. Bazilevs, L. Beirão da Veiga, J. A. Cottrell, T. J. R. Hughes, and G. Sangalli. Isogeometric analysis: Approximation, stability and error estimates for h-refined meshes. *Mathematical Methods and Models in Applied Sciences*, 16:1031–1090, 2006.

[19] Y. Bazilevs, V. M. Calo, J. A. Cottrell, J. Evans, T. J. R. Hughes, S. Lipton, M. A. Scott, and T. W. Sederberg. Isogeometric analysis using T-splines. *Computer Methods in Applied Mechanics and Engineering*, 199:229–263, 2010.

[20] Y. Bazilevs, V. M. Calo, T. J. R. Hughes, and Y. Zhang. Isogeometric fluid-structure interaction: theory, algorithms, and computations. *Computational Mechanics*, 43:3–37, 2008.

[21] L. Beirão da Veiga, A. Buffa, D. Cho, and G. Sangalli. Isogeometric analysis using T-splines on two patch geometries. *Computer Methods in Applied Mechanics and Engineering*, 200:1787–1803, 2011.

[22] L. Beirão da Veiga, A. Buffa, J. Rivas, and G. Sangalli. Some estimates for h-p-k-refinement in isogeometric analysis. *Numerische Mathematik*, 2011.

[23] D. J. Benson, Y. Bazilevs, M. C. Hsu, and T. J. R. Hughes. Isogeometric shell analysis: The Reissner-Mindlin shell. *Computer Methods in Applied Mechanics and Engineering*, 199:276–289, 2009.

[24] M. Bern and P. Plassmann. *Mesh Generation*, chapter 6, pages 291–332. Handbook of Computational Geometry. Elsevier, 1999.

[25] M. J. Borden, M. A. Scott, J. Evans, and T. J. R. Hughes. Isogeometric finite element data structures based on Bézier extraction of NURBS. *International Journal for Numerical Methods in Engineering*, 2010.

[26] D. Braess. *Finite elements: theory, fast solvers, and applications in elasticity theory*. Cambridge University Press, 2007.

[27] S. Brenner and L. Scott. *The Mathematical Theory of Finite Element Methods*. Number 15 in Texts in Applied Mathematics. Springer, Berlin, Heidelberg, New York, second edition, 2002.

[28] A. Buffa, D. Cho, and G. Sangalli. Linear independence of the T-spline blending functions associated with some particular T-meshes. *Computer Methods in Applied Mechanics and Engineering*, 199:1437 – 1445, 2010.

[29] A. Buffa, G. Sangalli, and R. Vázquez. Isogeometric analysis in electromagnetics: B-splines approximation. *Computer Methods in Applied Mechanics and Engineering*, 199:1143–1152, 2010.

[30] C. Carstensen. Some remarks on the history and future of averaging techniques in a posteriori finite element error analysis. *ZAMM*, 84:3–21, 2004.

[31] P. G. Ciarlet. *Mathematical Elasticity: Three-Dimensional Elasticity*. Elsevier, 1994.

[32] P. G. Ciarlet. *The Finite Element Method for Elliptic Problems*. SIAM, 2002.

[33] P. G. Ciarlet and P. A. Raviart. Interpolation theory over curved elements, with applications to finite element methods. *Computer methods in applied mechanics and engineering*, 1:217–249, 1972.

[34] E. Cohen, T. Martin, R. M. Kirby, T. Lyche, and R. F. Riesenfeld. Analysis-aware modeling: Understanding quality considerations in modeling for isogeometric analysis. *Computer Methods in Applied Mechanics and Engineering*, 199:334–356, 2010.

[35] J. A. Cottrell, T. J. R. Hughes, and Y. Bazilevs. *Isogeometric Analysis: Toward Integration of CAD and FEA*. John Wiley & Sons, 2009.

[36] J. A. Cottrell, A. Reali, Y. Bazilevs, and T. J. R. Hughes. Isogeometric analysis of structural vibrations. *Computer Methods in Applied Mechanics and Engineering*, 195:5257–5296, 2006.

[37] T. A. Davis. *Direct Methods for Sparse Linear Systems*. SIAM, 2006.

[38] M. de Berg, M. van Kreveld, M. Overmars, and O. Schwarzkopf. *Computational Geometry: Algorithms and Applications*. Springer, 1997.

[39] C. de Boor. *A Practical Guide to Splines*. Springer, 2001.

[40] J. Deng, F. Chen, X. Li, C. Hu, W. Tong, Z. Yang, and Y. Feng. Polynomial splines over hierarchical T-meshes. *Graphical Models*, 70:76–86, 2008.

[41] J. Donea and A. Huerta. *Finite Element Methods for Flow Problems*. John Wiley and Sons, 2003.

[42] M. R. Dörfel, B. Jüttler, and B. Simeon. Adaptive isogeometric analysis by local h-refinement with T-splines. *Computer Methods in Applied Mechanics and Engineering*, 199:264–275, 2010.

[43] R. Echter and M. Bischoff. Numerical efficiency, locking and unlocking of NURBS finite elements. *Computer Methods in Applied Mechanics and Engineering*, 199:374–382, 2010.

[44] J. A. Evans, Y. Bazilevs, I. Babuška, and T. J. R. Hughes. N-widths, sup-infs, and optimality ratio for the k-version of the isogeometric finite element method. *Computer Methods in Applied Mechanics and Engineering*, 198:1726–1741, 2009.

[45] G. Farin. *NURBS: from projective geometry to practical use.* A.K. Peters, 1999.

[46] G. Farin. *Curves and Surfaces for CAGD: A Practical Guide.* Morgan Kaufmann, 5th edition, 2002.

[47] D. R. Forsey and R. H. Bartels. Hierarchical B-spline refinement. *Computer Graphics*, 22:205–212, 1988.

[48] R. W. Freund, G. H. Golub, and N. M. Nachtigal. Iterative solution of linear systems. *Acta Numerica*, 1:57–100, 1992.

[49] D. Fußeder. Adaptive simulation and error estimation for isogeometric finite elements. Diplomarbeit, Technische Universität München, 2011.

[50] O. Gonzalez and A. M. Stuart. *A First Course in Continuum Mechanics.* Cambridge University Press, 2008.

[51] P. L. Gould. *Introduction to Linear Elasticity.* Springer, 1994.

[52] E. Grinspun, P. Krysl, and P. Schröder. CHARMS: A simple framework for adaptive simulation. *ACM Transactions on Graphics*, pages 281–290, 2002.

[53] C. Grossmann, H.-G. Roos, and M. Stynes. *Numerical Treatment of Partial Differential Equations.* Springer, 2007.

[54] C. Heinrich, B. Simeon, and S. Boschert. A finite volume method on NURBS geometries and its application in isogeometric fluid-structure interaction. (submitted), 2011.

[55] K. Höllig. *Finite Element Methods with B-Splines.* SIAM, 2003.

[56] J. Hoschek and D. Lasser. *Fundamentals of Computer Aided Geometric Design.* AK Peters, Wellesley, MA, USA, 1993.

[57] T. J. R. Hughes. *The Finite Element Method.* Dover Publ., Mineola, New York, 2000.

[58] T. J. R. Hughes, J. A. Cottrell, and Y. Bazilevs. Isogeometric analysis: CAD, finite elements, NURBS, exact geometry and mesh refinement. *Computer Methods in Applied Mechanics and Engineering*, 194:4135–4195, 2005.

[59] T. J. R. Hughes, A. Reali, and G. Sangalli. Efficient quadrature for NURBS-based isogeometric analysis. *Computer Methods in Applied Mechanics and Engineering*, 199:301–313, 2010.

[60] P. Kagan, A. Fischer, and P. Z. Bar-Yoseph. New B-spline finite element approach for geometrical design and mechanical analysis. *International Journal for Numerical Methods in Engineering*, 41:425–458, 1998.

[61] P. Kagan, A. Fischer, and P. Z. Bar-Yoseph. Mechanically bases models: Adaptive refinement for B-spline finite element. *International Journal for Numerical Methods in Engineering*, 57:1145–1175, 2003.

[62] J. Kiendl, Y. Bazilevs, M.-C. Hsu, R. Wüchner, and K.-U. Bletzinger. The bending strip method for isogeometric analysis of Kirchhoff-Love shell structures comprised of multiple patches. *Computer Methods in Applied Mechanics and Engineering*, 199:2403–2416, 2010.

[63] J. Kiendl, K.-U. Bletzinger, J. Linhard, and J. Wüchner. Isogeometric shell analysis with Kirchhoff-Love elements. *Computer Methods in Applied Mechanics and Engineering*, 198:3902–3914, 2009.

[64] R. Kraft. Adaptive and linearly independent multilevel B-splines. In A. Le Méhauté, C. Rabut, and L. L. Schumaker, editors, *Surface Fitting and Mulitresolution Methods*, pages 209–218. Vanderbilt University Press, Nashville, 1997.

[65] R. Kraft. *Adaptive und Linear Unabhängige Multilevel B-Splines und ihre Anwendungen*. PhD thesis, Universität Stuttgart, 1998.

[66] P. Krysl, E. Grinspun, and P. Schröder. Natural hierarchical refinement for finite element methods. *International Journal for Numerical Methods in Engineering*, 56:1109–1124, 2003.

[67] M. Lenoir. Optimal isoparametric finite elements and error estimates for domains involving curved boundaries. *SIAM Journal on Numerical Analysis*, 23:562–580, 1986.

[68] X. Li, J. Zheng, T. W. Sederberg, T. J. R. Hughes, and M. A. Scott. On linear independence of T-splines. Technical Report 10-40, ICES, 2010.

[69] S. Lipton, J. A. Evans, Y. Bazilevs, T. Elguedj, and T. J. R. Hughes. Robustness of isogeometric structural discretizations under severe mesh distortion. *Computer Methods in Applied Mechanics and Engineering*, 199:357–373, 2010.

[70] N. G. Manh, A. Evgrafov, A. R. Gersborg, and J. Gravesen. Isogeometric shape optimization of vibrating membranes. *Computer Methods in Applied Mechanics and Engineering*, 200:1343–1353, 2011.

[71] C. Manni, F. Pelosi, and M. L. Sampoli. Generalized B-splines as a tool in isogeometric analysis. *Computer Methods in Applied Mechanics and Engineering*, 200:867–881, 2011.

[72] J. E. Marsden and T. J. R. Hughes. *Mathematical Foundations of Elasticity*. Dover Publications, New York, 1994.

[73] J. M. Melenk and I. Babuška. The partion of unity finite element method: Basic theory and applications. *Computer methods in applied mechanics and engineering*, 139:289 – 314, 1996.

[74] N. Nguyen-Thanh, H. Nguyen-Xuan, S. P. A. Bordas, and T. Rabczuk. Isogeometric analysis using polynomial splines over hierarchical T-meshes for two-dimensional elastic solids. *Computer Methods in Applied Mechanics and Engineering*, 200:1892–1908, 2011.

[75] R. W. Ogden. *Non-linear elastic deformations*. Dover Publications, 1997.

[76] L. Piegl and W. Tiller. *The NURBS Book*. Monographs in Visual Communication. Springer, New York, 2nd edition, 1997.

[77] H. Prautzsch, W. Boehm, and M. Paluszny. *Bézier and B-Spline Techniques*. Springer, Berlin, Heidelberg, New York, 2002.

[78] D. Schillinger and E. Rank. An unfitted hp-adaptive finite element method based on hierarchical B-splines for interface problems of complex geometry. *Computer Methods in Applied Mechanics and Engineering*, 200:3358–3380, 2011.

[79] A. Schmidt and K. G. Siebert. *Design of adaptive finite element software: the finite element toolbox ALBERTA*. Springer, 2005.

[80] L. Schumaker. *Spline Functions: Basic Theory*. Cambridge University Press, 2007.

[81] C. Schwab. *p- and hp- Finite Element Methods: Theory and applications in Solid and Fluid Mechanics*. Oxford University Press, 1998.

[82] M. A. Scott, M. J. Borden, C. V. Verhoosel, T. W. Sederberg, and T. J. R. Hughes. Isogeometric finite element data structures based on Bézier extraction of T-splines. Technical Report 10-45, ICES, 2010.

[83] M. A. Scott, X. Li, T. W. Sederberg, and T. J. R. Hughes. Local refinement of analysis-suitable T-splines. Technical Report 11-06, ICES, 2011.

[84] T. W. Sederberg, D. L. Cardon, G. T. Finnigan, N. S. North, J. Zheng, and T. Lyche. T-spline simplification and local refinement. *ACM Transactions on Graphics*, 23:276 – 283, 2004.

[85] T. W. Sederberg, J. Zheng, A. Bakenov, and A. Nasri. T-splines and T-NURCCS. *ACM Transactions on Graphics*, 22:477–484, 2003.

[86] R. Sevilla, S. Férnandes-Méndez, and A. Huerta. NURBS-enhanced finite element method (NEFEM). *International Journal for Numerical Methods in Engineering*, 76:56–83, 2008.

[87] B. Simeon and A.-V. Vuong. Identification and specification of benchmark problem with typical geometries, computation of reference solutions. EXCITING Report 3.1, Technische Universität München, 2009.

[88] B. Simeon and A.-V. Vuong. Isogeometric structural prototype solver. EXCIT-ING Report 3.2, Technische Universität München, 2009.

[89] A. H. Squillacote. *The ParaView Guide*. Kitware Inc., 2007.

[90] O. Steinbach. *Numerical Approximation Methods for Elliptic Boundary Value Problems*. Springer, 2008.

[91] B. Szabó, A. Düster, and E. Rank. *The p-version of the finite element method*, volume 1 of *Encyclopedia of Comupational Mechanics*, chapter 5. Wiley, New York, 2004.

[92] R. Teman and A. Miranville. *Mathematical Modeling in Continuum Mechanics*. Cambridge University Press, 2005.

[93] I. Temizer, P. Wriggers, and T. J. R. Hughes. Contact treatment in isogeometric analysis with NURBS. *Computer Methods in Applied Mechanics and Engineering*, 200:1100–1112, 2011.

[94] R. Verfürth. *A Review of A Posteriori Error Estimation and Adaptive Mesh-Refinement Techniques*. Wiley-Teubner, 1996.

[95] A.-V. Vuong, C. Giannelli, B. Jüttler, and B. Simeon. A hierarchical approach to local refinement in isogeometric analysis. *Computer Methods in Applied Mechanics and Engineering*, 200:pp. 3554–3567, 2011.

[96] A.-V. Vuong, C. Heinrich, and B. Simeon. ISOGAT: A 2D tutorial Matlab code for isogeometric analysis. *Computer Aided Geometric Design*, 27:644–655, 2010.

[97] A.-V. Vuong and B. Simeon. On isogeometric analysis and its usage for stress calculation. In H. G. Bock, X. P. Hoang, R. Rannacher, and J. P. Schlöder, editors, *Modeling, Simulation and Optimization of Complex Processes*, pages 305 – 314. Springer, 2012.

[98] W. A. Wall, M. A. Frenzel, and C. Cyron. Isogeometric stuctural shape optimization. *Computer Methods in Applied Mechanics and Engineering*, 197:2976–2988, 2008.

[99] G. Xu, B. Mourrain, R. Duvigneau, and A. Galligo. Optimal analysis-aware parameterization of computational domain in isogeometric analysis. In *Proc. of Geometric Modeling and Processing (GMP 2010)*, pages 236–254, 2010.

[100] C. Yassouridis. Isogeometric modelling and simulation in railway technology. Diplomarbeit, Technische Universität München, 2010.

[101] A. Ženíšek. A general theorem on triangular finite $C^{(m)}$-elements. *RAIRO*, R-2:119–127, 1974.

[102] O. C. Zienkiewicz and J. Z. Zhu. A simple error estimator and adaptive procedure for practical engineering analysis. *International Journal for Numerical Methods in Engineering*, 24:337–357, 1987.